Textile Mill Management: Theory and Practice

Textile Mill Management: Theory and Practice

Naresh Grover

RANDOM

RANDOM PUBLICATIONS
NEW DELHI (INDIA)

Textile Mill Management: Theory and Practice

ISBN 978-93-5111-873-2

Published in 2016 in India by

RANDOM PUBLICATIONS

4376-A/4B, Gali Murari Lal, Ansari Road
New Delhi-110 002
Phone : +9111-43580356, 011-23289044, 011-43142548
e-mail: sales@randompublications.com,
info@randompublications.com, randomexports@gmail.com

Reprinted 2026

Type Setting by : Friends Media, Delhi-110089

Preface

Cotton mills get the cotton shipped to them in large, 500 pound bales. When the cotton comes out of a bale, it is all packed together and still contains vegetable matter. The bale is broken open using a machine with large spikes. It is called an Opener. In order to fluff up the cotton and remove the vegetable matter, the cotton is sent through a picker, or similar machines. The cotton is fed into a machine known as a picker, and gets beaten with a beater bar in order to loosen it up. It is fed through various rollers, which serve to remove the vegetable matter. The cotton, aided by fans, then collects on a screen and gets fed through more rollers till it emerges as a continuous soft fleecy sheet, known as a lap.

The main challenge for the textile industry today is to modify production methods, so they are more ecologically friendly at a competitive price, by using safer dyes and chemicals and by reducing cost of effluent treatment/disposal. Recycling has become a necessary element, not because of the shortage of any item, but because of the need to control pollution.

Textile manufacturing begins with fiber, which can be harvested from natural resources (e.g., cotton, wool, silk, or ramie), manufactured from cellulosic materials (e.g., rayon or acetate), or made of man-made synthetic materials (e.g., polyester, nylon, or acrylic). After the raw fibers are shipped from the farm or the chemical plant, they pass through four main stages of processing.

The present book is primarily meant for the post-graduate students and researchers of economics, management and labour studies at the university level. It is a comprehensive one that deals with the crucial issues of industrial labour, wage economics, managerial aspects, social and legal aspects of the giant, century old cotton textile mills industry.

– Author

Contents

1

Textile Manufacturing

Textile manufacturing is a major industry. It is based on the conversion of fibre into yarn, yarn into fabric. These are then dyed or printed, fabricated into clothes. Different types of fiber are used to produce yarn. Cotton remains the most important natural fibre, so is treated in depth. There are many variable processes available at the spinning and fabric-forming stages coupled with the complexities of the finishing and colouration processes to the production of a wide ranges of products. There remains a large industry that uses hand techniques to achieve the same results.

PROCESSING OF COTTON

Cotton is the world's most important natural fibre. In the year 2007, the global yield was 25 million tons from 35 million hectares cultivated in more than 50 countries.

There are six stages

- Cultivating and Harvesting
- Preparatory Processes
- Spinning
- Weaving or Knitting
- Finishing
- Marketing

Cultivating and harvesting

Cotton is grown anywhere with long, hot dry summers with plenty of sunshine and low humidity. Indian cotton, gossypium arboreum, is finer but the staple is only suitable for hand processing. American cotton, gossypium hirsutum, produces the longer staple needed for machine production. Planting is from September to mid November and the crop is harvested between March and June. The cotton bolls are harvested by stripper harvesters and spindle pickers, that remove the entire boll from the plant. The cotton boll is the seed pod of the cotton plant, attached to each of the thousands of seeds are fibres about 2.5 cm long.

Ginning

The seed cotton goes in to a Cotton gin. The cotton gin separates seeds and removes the "trash" (dirt, stems and leaves) from the fibre. In a saw gin, circular saws grab the fibre and pull it through a grating that is too narrow for the seeds to pass. A roller gin is used with longer staple cotton. Here a leather roller captures the cotton. A knife blade, set close to the roller, detaches the seeds by drawing them through teeth in circular saws and revolving brushes which clean them away.

The ginned cotton fibre, known as lint, is then compressed into bales which are about 1.5 m tall and weigh almost 220 kg. Only 33% of the crop is usable lint. Commercial cotton is priced by quality, and that broadly relates to the average length of the staple, and the variety of the plant. Longer staple cotton (2½ in to 1¼ in) is called Egyptian, medium staple (1¼ in to ¾ in) is called American upland and short staple (less than ¾ in) is called Indian.

The cotton seed is pressed into a cooking oil. The husks and meal are processed into animal feed, and the stems into paper.

Preparatory processes - preparation of yarn

- Ginning, bale-making and transportation is done in the country of origin.
- Opening and cleaning

Fig. Platt Bros. Picker

Cotton mills get the cotton shipped to them in large, 500 pound bales. When the cotton comes out of a bale, it is all packed together and still contains vegetable matter. The bale is broken open using a machine with large spikes. It is called an Opener. In order to fluff up the cotton and remove the vegetable matter, the cotton is sent through a picker, or similar machines. The cotton is fed into a machine known as a picker, and gets beaten with a beater bar in order to loosen it up. It is fed through various rollers, which serve to remove the vegetable matter. The cotton, aided by fans, then collects on a screen and gets fed through more rollers till it emerges as a continuous soft fleecy sheet, known as a lap.

Blending,

Scutching refers to the process of cleaning cotton of its seeds and other impurities. The first scutching machine was invented in 1797, but did not come into further mainstream use until after 1808 or 1809, when it was introduced and used in Manchester, England. By 1816, it had become generally adopted. The scutching machine worked by passing the cotton through a pair of rollers, and then striking it with iron or steel bars called beater bars or beaters. The beaters, which turn very quickly, strike the cotton hard and knock the seeds out. This process is done over a series of parallel bars so as to allow the seeds to fall through. At the same time, air is blown across the bars, which carries the cotton into a cotton chamber.

Carding

Fig. Carding machine

Carding: the fibres are separated and then assembled into a loose strand (sliver or tow) at the conclusion of this stage.

The cotton comes off of the picking machine in laps, and is then taken to carding machines. The carders line up the fibres nicely to make them easier to spin. The carding machine consists mainly of one big roller with smaller ones surrounding it. All of the rollers are covered in small teeth, and as the cotton progresses further on the teeth get finer (i.e. closer together). The cotton leaves the carding machine in the form of a sliver; a large rope of fibres.

Fig. A Combing machine

Note: In a wider sense Carding can refer to these four processes: Willowing- loosening the fibres; Lapping- removing the dust to create a flat sheet or lap of cotton; Carding- combing the tangled lap into a thick rope of 1/2 in in diameter, a sliver; and Drawing- where a drawing frame combines 4 slivers into one- repeated for increased quality.

- Combing is optional, but is used to remove the shorter fibres, creating a stronger yarn.
- Drawing the fibres are straightened

Several slivers are combined. Each sliver will have thin and thick spots, and by combining several slivers together a more consistent size can be reached. Since combining several slivers produces a very thick rope of cotton fibres, directly after being combined the slivers are separated into rovings. These rovings (or slubbings) are then what are used in the spinning process. Generally speaking, for machine processing, a roving is about the width of a pencil.

- Drawing frame: Draws the strand out
- Slubbing Frame: adds twist, and winds onto bobbins
- Intermediate Frames: are used to repeat the slubbing process to produce a finer yarn.
- Roving frames: reduces to a finer thread, gives more twist, makes more regular and even in thickness, and winds onto a smaller tube.

SPINNING - YARN MANUFACTURE

Spinning

Most spinning today is done using Break or Open-end spinning, this is a technique where the staples are blown by air into a rotating drum, where they attach themselves to the tail of formed yarn that is continually being drawn out of the chamber. Other methods of break spinning use needles and electrostatic forces. This method has replaced the older methods of ring and mule spinning. It is also is easily adapted for artificial fibres.

The spinning machines takes the roving, thins it and twists it, creating yarn which it winds onto a bobbin.

In mule spinning the roving is pulled off a bobbin and fed through some rollers, which are feeding at several different speeds. This thins the roving at a consistent rate. If the roving was not a consistent size, then this step could cause a break in the yarn, or could jam the machine. The yarn is twisted through the spinning of the bobbin as the carriage moves out, and is rolled onto a cylinder called a spindle, which then produces a cone-shaped bundle of fibres known as a "cop", as the carriage returns. Mule spinning produces a finer thread than the less skilled ring spinning.

- The mule was an intermittent process, as the frame advanced and returned a distance of 5ft.It was the descendant of 1779 Crompton device. It produces a softer less twisted thread that was favoured for fines and for weft.
- The ring was a descendant of the Arkwright water Frame 1769. It was a continuous process, the yarn was coarser, had a greater twist and was stronger so was suited to be warp. Ring spinning is slow due to the distance the thread must pass around the ring, other methods have been introduced.

Sewing thread, was made of several threads twisted together, or doubled.

Checking

This is the process where each of the bobbins is rewound to give a tighter bobbin.

Folding and twisting

Plying is done by pulling yarn from two or more bobbins and twisting it together, in the opposite direction that in which it was spun. Depending on the weight desired, the cotton may or may not be plied, and the number of strands twisted together varies.

Gassing

Gassing is the process of passing yarn, as distinct from fabric very rapidly through a series of Bunsen gas flames in a gassing frame, in order to burn off the projecting fibres and make the thread round and smooth and also brighter. Only the better qualities of yarn are gassed, such as that used for voiles, poplins, venetians, gabardines, many Egyptian qualities, etc. There is a loss of weight in gassing, which varies' about 5 to 8 per cent., so that if a 2/60's yarn is required

2/56's would be used. The gassed yarn is darker in shade afterwards, but should not be scorched.

Fig. Mule spinning

Fig. Mule spinning

Fig. Ring spinning

Fig. Ring spinning

Measurements

- Cotton Counts: The number of pieces of thread, 840 yards long needed to make up 1 lb weight. 10 count cotton means that 10x840 yd weighs 1 lb. This is coarser than 40 count cotton where 40x840 yards are needed. In the United Kingdom, Counts to 40s are coarse (Oldham Counts), 40 to 80s are medium counts and above 80 is a fine count. In the United States ones to 20s are coarse counts.
- Hank: A length of 7 leas or 840 yards
- Thread: A length of 54 in (the circumference of a warp beam)
- Bundle: Usually 10 lb
- Lea: A length of 80 threads or 120 yards
- Denier: this is an alternative method. It is defined as a number that is equivalent to the weight in grams of 9000m of a single yarn. 15 denier is finer than 30 denier.
- Tex: is the weight in grams of 1 km of yarn.

The worsted hank is only 560 yd

WEAVING-FABRIC MANUFACTURE

The weaving process uses a loom. The lengthway threads are known as the warp, and the cross way threads are known as the weft. The warp which must be strong needs to be presented to loom on a warp beam. The weft passes across the loom in a shuttle, that carries the yarn on a pirn. These pirns are automatically changed by the loom. Thus, the yarn needs to be wrapped onto a beam, and onto pirns before weaving can commence.

Winding

After being spun and plied, the cotton thread is taken to a warping room where the winding machine takes the required length of yarn and winds it onto warpers bobbins

Warping or beaming

Fig. A Warper

Racks of bobbins are set up to hold the thread while it is rolled onto the warp bar of a loom. Because the thread is fine, often three of these would be combined to get the desired thread count.

Sizing

Slasher sizing machine needed for strengthening the warp by adding starch to reduce breakage of the yarns.

Drawing in, Looming

The process of drawing each end of the warp separately through the dents of the reed and the eyes of the healds, in the order indicated by the draft.

Pirning (Processing the weft)

Pirn winding frame was used to transfer the weft from cheeses of yarn onto the pirns that would fit into the shuttle

Weaving

At this point, the thread is woven. Depending on the era, one person could manage anywhere from 3 to 100 machines. In the mid nineteenth century, four was the standard number. A skilled weaver in 1925 would run 6 Lancashire Looms. As time progressed new mechanisms were added that stopped the loom any time something went wrong.

The mechanisms checked for such things as a broken warp thread, broken weft thread, the shuttle going straight across, and if the shuttle was empty. Forty of these Northrop Looms or automatic looms could be operated by one skilled worker.

Fig. A Draper loom in textile museum, Lowell, Massachusetts

The three primary movements of a loom are shedding, picking, and beating-up.

- Shedding: The operation of dividing the warp into two lines, so that the shuttle can pass between these lines. There are two general kinds of sheds-"open" and "closed." Open Shed-The warp threads are moved when the pattern requires it-from one line to the other. Closed Shed-The warp threads are all placed level in one line after each pick.
- Picking:The operation of projecting the shuttle from side to side of the loom through the division in the warp threads. This is done by the overpick or underpick motions. The overpick is suitable for quick-running looms, whereas the underpick is best for heavy or slow looms.
- Beating-up: The third primary movement of the loom when making cloth, and is the action of the reed as it drives each pick of weft to the fell of the cloth.

The Lancashire Loom was the first semi-automatic loom. Jacquard looms and Dobby looms are looms that have sophisticated methods of shedding. They may be separate looms, or mechanisms added to a plain loom. A Northrop Loom was fully automatic and was mass produced between 1909 and the mid-1960s. Modern looms run faster and do not use a shuttle: there are air jet looms, water jet looms and rapier looms.

Measurements

- Ends and Picks: Picks refer to the weft, ends refer to the warp. The coarseness of the cloth can be expressed as the number of picks and ends per quarter inch square, or per inch square. Ends is always written first. For example: Heavy domestics are made from coarse yarns, such as 10's to 14's warp and weft, and about 48 ends and 52 picks.

Associated job titles

- Piecer
- Scavenger
- Weaver
- Tackler
- Draw boy
- Pirner

Issues

When a hand loom was located in the home, children helped with the weaving process from an early age. Piecing needs dexterity, and a child can be as productive as an adult. When weaving moves from the home to the mill, children are often allowed to help their older sisters, and laws have to be made to prevent child labour becoming established.

KNITTING- FABRIC MANUFACTURE

Fig. A circular knitting machine.

Fig. Close-up on the needles.

Knitting by machine is done in two different ways; warp and weft. Weft knitting (as seen in the pictures) is similar in method to hand knitting with stitches all connected to each other horizontally. Various weft machines can be configured to produce textiles from a single spool ofyarn or multiple spools depending on the size of the machine cylinder (where the needles are bedded). In a warp knit there are many pieces of yarn and there are vertical chains, zigzagged together by crossing the yarn. Cotton

Warp knits do not stretch as much as a weft knit, and it is run-resistant. A weft knit is not run-resistant, but stretches more. This is especially true if spools of spandex are processed from separate spool containers and interwoven through the cylinder with cotton yarn, giving the finished product more flexibility and making it less prone to having a 'baggy' appearance. The average t-shirt is a weft knit.

FINISHING- PROCESSING OF TEXTILES

The woven cotton fabric in its loom-state, not only contains impurities, including warp size, but requires further treatment in order to develop its full textile potential. Furthermore, it may receive considerable added value by applying one or more finishing processes.

Desizing

Depending on the size that has been used, the cloth may be steeped in a dilute acid and then rinsed, or enzymes may be used to break down the size.

Scouring

Scouring, is a chemical washing process carried out on cotton fabric to remove natural wax and non-fibrous impurities (e.g. the remains of seed fragments) from the fibres and any added soiling or dirt. Scouring is usually carried in iron vessels called kiers. The fabric is boiled in an alkali, which forms a soap with free fatty acids (saponification). A kier is usually enclosed, so the solution of sodium hydroxide can be boiled under pressure, excluding oxygen which would degrade the cellulose in the fibre. If the appropriate reagents are used, scouring will also remove size from the fabric although desizing often precedes scouring and is considered to be a separate process known as fabric preparation. Preparation and scouring are prerequisites to most of the other finishing processes. At this stage even the most naturally white cotton fibres are yellowish, and bleaching, the next process, is required.

Bleaching

Bleaching improves whiteness by removing natural coloration and remaining trace impurities from the cotton; the degree of bleaching necessary is determined by the required whiteness and absorbency. Cotton being a vegetable fibre will be bleached using an oxidizing agent, such as dilute sodium hypochlorite or dilute hydrogen peroxide. If the fabric is to be dyed a deep shade, then lower levels of bleaching are acceptable, for example. However, for white bed sheetings and medical applications, the highest levels of whiteness and absorbency are essential.

Mercerising

A further possibility is mercerizing during which the fabric is treated with caustic soda solution to cause swelling of the fibres. This results in improved lustre, strength and dye affinity. Cotton is mercerized under tension, and all alkali must be washed out before the tension is released or shrinkage will take place. Mercerizing can take place directly on grey cloth, or after bleaching.

Many other chemical treatments may be applied to cotton fabrics to produce low flammability, crease resist and other special effects but four important non-chemical finishing treatments are:

Singeing

Singeing is designed to burn off the surface fibres from the fabric to produce smoothness. The fabric passes over brushes to raise the fibres, then passes over a plate heated by gas flames.

Raising

Another finishing process is raising. During raising, the fabric surface is treated with sharp teeth to lift the surface fibres, thereby imparting hairiness, softness and warmth, as in flannelette.

Calendering

Calendering is the third important mechanical process, in which the fabric is passed between heated rollers to generate smooth, polished or embossed effects depending on roller surface properties and relative speeds.

Shrinking (Sanforizing)

Finally, mechanical shrinking (sometimes referred to as sanforizing), whereby the fabric is forced to shrink width and/or lengthwise, creates a fabric in which any residual tendency to shrink after subsequent laundering is minimal.

Dyeing

Finally, cotton is an absorbent fibre which responds readily to colouration processes. Dyeing, for instance, is commonly carried out with an anionic direct dye by completely immersing the fabric (or yarn) in an aqueous dyebath according to a prescribed procedure. For improved fastness to washing, rubbing and light, other dyes such as vats and reactives are commonly used. These require more complex chemistry during processing and are thus more expensive to apply.

Printing

Printing, on the other hand, is the application of colour in the form of a paste or ink to the surface of a fabric, in a predetermined pattern. It may be considered as localised dyeing. Printing designs onto already dyed fabric is also possible.

ECONOMIC, ENVIRONMENTAL AND POLITICAL CONSEQUENCES OF COTTON MANUFACTURE

The growth of cotton is divided into two segments i.e. organic and genetically modified. Cotton crop provides livelihood to millions of people but its production is becoming expensive because of high water consumption, use of expensive pesticides, insecticides and fertiliser. Genetically Modified products aim to increase disease resistance and reduce the water required. The organic sector was worth $583 million. Genetically Modified cotton, in 2007, occupied 43% of cotton growing areas.

Cotton is farmed intensively and uses large amounts of fertilizer and 25% of the world's insecticides. Native Indian varieties of cotton were rainwater fed, but modern hybrids used for the mills need irrigation, which spreads pests. The 5% of cotton-bearing land in India uses 55% of all pesticides used in India.

The consumption of energy in form of water and electricity is relatively high, especially in processes like washing, de-sizing, bleaching, rinsing, dyeing, printing, coating and finishing. Processing is time consuming. The major portion of water in textile industry is used for wet processing of textile (70 per cent). Approximately 25 per cent of energy in the total textile production like fibre production, spinning, twisting, weaving, knitting, clothing manufacturing etc. is used in dyeing. About 34 per cent of energy is consumed in spinning, 23 per cent in weaving, 38 per cent in chemical wet processing and five per cent in miscellaneous processes. Power dominates consumption pattern in spinning and weaving, while thermal energy is the major factor for chemical wet processing. Before mechanisation, cotton was harvested manually by farmers in India and by African slaves in America. In 2012 Uzbekistan was a major exporter of cotton and uses manual labour during the harvest. Human rights groups claim that health care professionals and children are forced to pick cotton.

PROCESSING OF OTHER VEGETABLE FIBRES

FLAX

Flax is a bast fibre, which means it comes in bundles under the bark of the Linum usitatissimum plant. The plant flowers and is harvested.

- Retting
- Breaking
- Scutching
- Hackling or combing

It is now treated like cotton.

JUTE

Jute is a bast fibre, which comes from the inner bark of the plants of the Corchorus genus. It is retted like flax, sundried and baled. When spinning a small amount of oil must be added to the fibre. It can be bleached and dyed. It was used for sacks and bags but is now used for the backing for carpets. Jute can be blended with other fibres to make composite fabrics and work continues in Bangladesh to refine the processes and extend the range of usage possible. In the 1970s, jute-cotton composite fabrics were known asjutton fabrics.

HEMP

Hemp is a bast fibre from the inner bark of Cannabis sativa. It is difficult to bleach, it is used for making cord and rope.

- Retting
- Separating
- Pounding

Other bast fibres

These bast fibres can also be used: kenaf, urena, ramie, nettle.

Other leaf fibres

Sisal is the main leaf fibre used; others are: abacá and henequen.

PROCESSING OF PROTEIN FIBRES

WOOL

Wool comes from domesticated sheep. It forms two products, woolens and worsteds. The sheep has two sorts of wool and it is the inner coat that is used. This can be mixed with wool that has been recovered from rags. Shoddy is the term for recovered wool that is not matted, while mungo comes from felted wool. Extract is recovered chemically from mixed cotton/wool fabrics.

The fleece is cut in one piece from the sheep.This is then skirted to remove the soiled wool, and baled. It is graded into long wool where the fibres can be up to 15 in, but anything over 2.5 inches is suitable for combing into worsteds. Fibres less than that form short wool and are described as clothing or carding wool. At the mill the wool is scoured in a detergent to remove grease (the yolk) and impurities. This is done mechanically in the opening machine.

Vegetable matter can be removed chemically using sulphuric acid (carbonising). Washing uses a solution of soap and sodium carbonate. The wool is oiled before carding or combing.

- Woollens: Use noils from the worsted combs, mungo and shoddy and new short wool
- Worsteds

Combing: Oiled slivers are wound into laps, and placed in the circular comber. The worsted yarn gathers together to form a top. The shorter fibres or noils remain behind and are removed with a knife.

Silk

The processes in silk production are similar to those of cotton but take account that reeled silk is a continuous fibre. The terms used are different.

- Opening bales. Assorting skeins: where silk is sorted by colour, size and quality, scouring: where the silk is washed in water of 40 degrees for 12 hours to remove the natural gum, drying: either by steam heating or centrifuge, softening: by rubbing to remove any remaining hard spots.
- Silk throwing (winding). The skeins are placed on a reel in a frame with many others. The silk is wound onto spools or bobbins.
- Doubling and twisting. The silk is far too fine to be woven, so now it is doubled and twisted to make the warp, known as organzine and the weft, known as tram. In organzine each single is given a few twists per inch (tpi), and combine with several other singles counter twisted hard at 10 to 14 tpi. In tram the two singles are doubled with each other with a light twist, 3 to 6 tpi. Sewing thread is two tram threads, hard twisted, and machine-twist is made of three hard-twisted tram threads. Tram for the crepe process is twisted at up to 80 tpi to make it 'kick up'.
- Stretching. The thread is tested for consistent size. Any uneven thickness is stretched out. The resulting thread is reeled into containing 500 yd to 2500 yd. The skeins are about 50 in in loop length.
- Dyeing: the skeins are scoured again, and discoloration removed with a sulphur process. This weakens the silk. The skeins are now tinted or dyed. They are dried and rewound onto bobbins, spools and skeins. Looming, and the weaving process on power looms is the same as with cotton.
- Weaving. The organzine is now warped. This is a similar process to in cotton. Firstly, thirty threads or so are wound onto a warping reel, and then using the warping reels, the threads are beamed. A thick layer of paper is laid between each layer on the beam to stop entangling.

PROCESSING OF SYNTHETIC FIBRES

DISCUSSION OF TYPES OF SYNTHETIC FIBERS

Synthetic fibers are the result of extensive development by scientists to improve upon the naturally occurring animal and plant fibers. In general, synthetic fibers are created by forcing, or extruding, fiber forming materials through holes (called spinnerets) into the air, thus forming a thread. Before synthetic fibers were developed, cellulose fibers were made from natural cellulose, which comes from plants.

The first artificial fiber, known as art silk from 1799 onwards, became known as viscose around 1894, and finally rayon in 1924. A similar product known as cellulose acetate was discovered in 1865. Rayon and acetate are both artificial fibers, but not truly synthetic, being made from wood. Although these artificial fibers were discovered in the mid-nineteenth century, successful modern manufacture began much later in the 1930s. Nylon, the first synthetic fiber, made its debut in the United States as a replacement for silk, and was used for parachutes and other military uses.

The techniques used to process these fibers in yarn are essentially the same as with natural fibers, modifications have to be made as these fibers are of great length, and have no texture such as the scales in cotton and wool that aid meshing.

TEXTILE MANUFACTURE DURING THE INDUSTRIAL REVOLUTION

Textile manufacture during the Industrial Revolution in Britain was centred in south Lancashire and the towns on both sides of thePennines. In Germany it was concentrated in the Wupper Valley, Ruhr Region and Upper Silesia, while in the United States it was in New England. The four key drivers of the Industrial Revolution were textile manufacturing, iron founding, steam power and cheap labour.

Before the 18th century, the manufacture of cloth was performed by individual workers, in the premises in which they lived and goods were transported around the country by packhorses or by river navigations and contour-following canals that had been constructed in the early 18th century. In the mid-18th century, artisans were inventing ways to become more productive. Silk, wool, and fustian fabrics were being eclipsed by cotton which became the most important textile.

Innovations in carding and spinning enabled by advances in cast iron technology resulted in the creation of larger spinning mules andwater frames. The machinery was housed in water-powered mills on streams. The need for more power stimulated the production of steam-powered beam engines, and rotative mill engines transmitting the power to line shafts on each floor of the mill. Surplus power capacity encouraged the construction of more sophisticated power looms working in weaving sheds. The scale of production in the mill towns round Manchester created a need for a commercial structure; for a cotton exchange and warehousing. The technology was used in woollen and worsted mills in the West Riding of Yorkshire and elsewhere.

ELEMENTS OF THE INDUSTRIAL REVOLUTION

Fig. The only surviving example of a Spinning Mule built by the inventor Samuel Crompton

The commencement of the Industrial Revolution is closely linked to a small number of innovations, made in the second half of the 18th century:

- Textiles – Cotton spinning using Richard Arkwright's water frame, James Hargreaves's Spinning Jenny, and Samuel Crompton'sSpinning Mule (a combination of the Spinning Jenny and the Water Frame). This was patented in 1769 and so came out of patent in 1783. The end of the patent was rapidly followed by the erection of many cotton mills. Similar technology was subsequently applied to spinning worsted yarn for various textiles and flax for linen.
- Steam power – The improved steam engine invented by James Watt and patented in 1775 was initially mainly used for pumping outmines, but from the 1780s was applied to power machines. This enabled rapid development of efficient semi-automated factories on a previously unimaginable scale in places where waterpower was not available.
- Iron founding – In the Iron industry, coke was finally applied to all stages of iron smelting, replacing charcoal. This had been achieved much earlier for lead and copper as well as for producing pig iron in a blast furnace, but the second stage in the production of bar iron depended on the use of potting and stamping (for which a patent expired in 1786) or puddling (patented by Henry Cort in 1783 and 1784).

These represent three 'leading sectors', in which there were key innovations, which allowed the economic take off by which the Industrial Revolution is usually defined. This is not to belittle many other inventions, particularly in the textile industry. Without earlier ones, such as the spinning jenny and flying shuttle in the textile industry and the smelting of pig iron with coke, these achievements might have been impossible. Later inventions such as the power loom and Richard Trevithick's high pressure steam engine were also important in the growing industrialisation of Britain. The application of steam engines to powering cotton mills and ironworks enabled these to be built in places that were most convenient because other resources were available, rather than where there was water to power a watermill.

PROCESSING OF COTTON

Cotton is the world's most important natural fibre. In the year 2007, the global yield was 25 million tons from 35 million hectares cultivated in more than 50 countries.

There are five stages

- Cultivating and Harvesting
- Preparatory Processes
- Spinning
- Weaving
- Finishing

INDUSTRY AND INVENTION

Before the 1760s, textile production was a cottage industry using mainlyflax and wool. A typical weaving family would own one hand loom, which would be operated by the man with help of a boy; the wife, girls and other women could make sufficient yarn for that loom. The knowledge of textile production had existed for centuries, and the manual methods had adequately provided enough cloth for the needs of that society. Cottonstarted to be imported and the balance of demand and supply was disturbed.

Two systems had developed for spinning: the simple wheel, which used anintermittent process and the more refined, Saxony wheel which drove a differential spindle and flyer with a heck that guided the thread onto the bobbin, as a continuous process. But neither of these wheels could produce enough thread for the looms after the invention by John Kay in 1734 of the flying shuttle. This made the loom twice as productive. Cloth production moved away from the cottage into manufactories. The first moves towards manufactories called mills were made in the spinning sector. The move in the weaving sector was later. By the 1820s, all cotton, wool and worsted was spun in mills; but this yarn went to outworking weavers who continued to work in their own homes. A mill that specialised in weaving fabric was called a weaving shed.

Early Inventions

During the second half of the 17th century, cotton goods were imported from India. In 1700, because of the competition with the wool and the linen industries, the government placed a ban on imported cotton goods. Cotton had become popular, however, and a home-based cotton industry sprung up using the raw material imported from the colonies. Since much of the imported cotton came from New England, ports on the west coast of Britain, such as Liverpool, Bristol and Glasgow, became important in determining the sites of the cotton industry. Of course, the wool and linen manufacturers made sure that many restrictions were imposed on the import of cotton, but, as cotton had become fashionable, there was little they could do to stop the trend.

Lancashire became a centre for the nascent cotton industry because the damp climate was better for spinning the yarn. As the cotton thread was not strong enough to use as warp, wool or linen or fustian had to be used. Lancashire was an existing wool centre.

The early advances in weaving had been halted by the lack of thread. The spinning process was slow and the weavers needed more cotton and wool thread than their families could produce. In the 1760s, James Hargreaves improved thread production when he invented the Spinning Jenny. By the end of the decade, Richard Arkwright had developed the water frame. This invention had two important consequences: it improved the quality of the thread, which meant that the cotton industry was no longer dependent on wool or linen to make the warp, and it took spinning away from the artisans' homes to specific locations where fast-flowing streams could provide the water power needed to drive the larger machines. The Western Pennines of Lancashire became the centre for the cotton industry. Not long after the invention of the water frame, Samuel Crompton combined the principals of the Spinning Jenny and the Water Frame to produce his Spinning Mule. This provided even tougher and finer cotton thread.

The textile industry was also to benefit from other developments of the period. As early as 1691, Thomas Savery had made a vacuum steam engine. His design, which was unsafe, was improved by Thomas Newcomen in 1698. In 1765, James Watt further modified Newcomen's engine to design an external condenser steam engine. Watt continued to make improvements on his design, producing a separate condenser engine in 1774 and a rotating separate condensing engine in 1781. Watt formed a partnership with a businessman called Matthew Boulton, and together they manufactured steam engines which could be used by industry.

Timeline of inventions

In 1734 in Bury, Lancashire, John Kay invented the flying shuttle — one of the first of a series of inventions associated with the cotton industry. The flying shuttle increased the width of cotton cloth and speed of production of a single weaver at a loom. Resistance by workers to the perceived threat to jobs delayed the widespread introduction of this technology, even though the higher rate of production generated an increased demand for spun cotton.

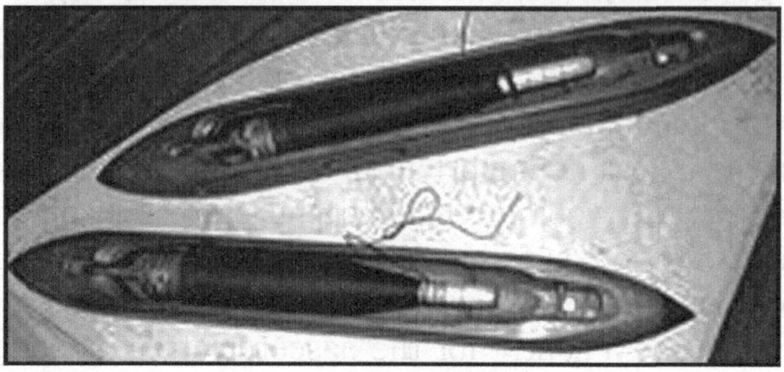

Fig. Shuttles

In 1738, Lewis Paul (one of the community of Huguenot weavers that had been driven out of France in a wave of religious persecution) settled in Birmingham and with John Wyatt, of that town, they patented the Roller Spinning machine and the flyer-and-bobbin system, for drawing wool to a more even thickness. Using two sets of rollers that travelled at different speeds yarn could be twisted and spun quickly and efficiently. This was later used in the first cotton spinning mill during the Industrial Revolution.

1742: Paul and Wyatt opened a mill in Birmingham which used their new rolling machine powered by donkey; this was not profitable and was soon closed. 1743: A factory opened in Northampton, fifty spindles turned on five of Paul and Wyatt's machines proving more successful than their first mill. This operated until 1764. 1748: Lewis Paul invented the hand driven carding machine. A coat of wire slips were placed around a card which was then wrapped around a cylinder. Lewis's invention was later developed and improved by Richard Arkwright and Samuel Crompton, although this came about under great suspicion after a fire at Daniel Bourn's factory in Leominsterwhich specifically used Paul and Wyatt's spindles. Bourn produced a similar patent in the same year. 1758: Paul and Wyatt based in Birmingham improved their roller spinning machine and took out a second patent. Richard Arkwright later used this as the model for his water frame.

Start of the Revolution

The Duke of Bridgewater's canal connected Manchester to the coal fields of Worsley. It was opened in July 1761. Matthew Boulton opened the Soho Foundry engineering works in Handsworth, Birmingham in 1762. These were both events that enabled cotton mill construction and the move away from home-based production. In 1764, Thorp Mill, the first water-powered cotton mill in the world was constructed at Royton, Lancashire, England. It was used for carding cotton. The multiple spindle spinning jenny was invented in 1764. James

Hargreaves is credited as the inventor. This machine increased the thread production capacity of a single worker — initially eightfold and subsequently much further. Others credit the original invention to Thomas Highs. Industrial unrest forced Hargreaves to leave Blackburn, but more importantly for him, his unpatented idea was exploited by others. He finally patented it in 1770. As a result, there were over 20,000 spinning jennies in use (mainly unlicensed) by the time of his death.

Richard Arkwright first spinning mill, Cromford Mill, Derbyshire, was built in 1771. It contained his invention the water frame. The water frame was developed from the spinning frame that Arkwright had developed with (a different) John Kay, from Warrington. The original design was again claimed by Thomas Highs: which he purposed he had patented in 1769. Arkwright used waterwheels to power the textile machinery. His initial attempts at driving the frame had used horse power, but a mill needed far more power. Using a waterwheel demanded a location with a ready supply of water, hence the mill at Cromford. This mill is preserved as part of the Derwent Valley MillsArkwright generated jobs and constructed accommodation for his workers which he moved into the area. This led to a sizeable industrial community. Arkwright protected his investment from industrial rivals and potentially disruptive workers. This model worked and he expanded his operations to other parts of the country.

Matthew Boulton partnership with Scottish engineer James Watt resulted, in 1775, in the commercial production of the more efficient Watt steam engine which used a separate condensor. Samuel Crompton of Bolton combined elements of the spinning jenny and water frame in 1779, creating the spinning mule. This mule produced a stronger thread than the water frame could. Thus in 1780, there were two viable hand operated spinning system that could be easily adapted to run by power of water. As early mules were suitable for producing yarn for use in the manufacture of muslin, and which were known as the muslin wheel or the Hall i' th' Wood (pronounced Hall-ith-wood) wheel. As with Kay and Hargreaves, Crompton was not able to exploit his invention for his own profit, and died a pauper.

In 1783 a mill was built in Manchester at Shudehill, at the highest point in the city away from the river. Shudehill Mill was powered by a 30 ft diameter waterwheel. Two storage ponds were built, and the water from one passed from one to the other turning the wheel. A steam driven pump returned the water to the higher reservoir. The steam engine was of the atmospheric type. An improvement devised by Joshua Wrigley, trialled in Chorlton-upon-Medlock used two Savery engines to supplement the river in driving onovershot waterwheel. In 1784, Edmund Cartwright invented the power loom, and produced a prototype in the following year. His initial venture to exploit this technology failed, although his advances were recognised by others in the industry. Others such as Robert Grimshaw (whose factory was destroyed in 1790 as part of the growing reaction against the mechanization of the industry) and Austin – developed the ideas further. In the 1790s industrialists, such as John Marshall at Marshall's Mill in Leeds, started to work on ways to apply some of the techniques which had proved so successful in cotton to other materials, such as flax. In 1803, William Radcliffe invented the dressing frame which was patented under the name of Thomas Johnson which enabled power looms to operate continuously.

Later developments

With the Cartwright Loom, the Spinning Mule and the Boulton & Watt steam engine, the pieces were in place to build a mechanised textile industry. From this point there were no new inventions, but a continuous improvement in technology as the mill-owner strove to reduce cost and improve quality. Developments in the transport infrastructure - the canals and, after 1831, the railways - facilitated the import of raw materials and export of finished cloth.

The use of water power to drive mills was supplemented by steam driven water pumps, and then superseded completely by the steam engines. For example, Samuel Gregjoined his uncle's firm of textile merchants, and, on taking over the company in 1782, he sought out a site to establish a mill. Quarry Bank Mill was built on the River Bollin atStyal in Cheshire. It was initially powered by a water wheel, but installed steam engines in 1810. In 1830, the average power of a mill engine was 48 hp, but Quarry Bank mill installed an new 100 hp water wheel. This was to change in 1836, when Horrocks & Nuttall, Preston took delivery of 160 hp double engine. William Fairbairn addressed the problem of line-shafting and was responsible for improving the efficiency of the mill. In 1815 he replaced the wooden turning shafts that drove the machines at 50rpm, to wrought iron shafting working at 250 rpm, these were a third of the weight of the previous ones and absorbed less power. The mill operated until 1959.

Fig. A Roberts loom in a weaving shed in 1835. Note the wrought iron shafting, fixed to the cast iron columns

In 1830, using an 1822 patent, Richard Roberts manufactured the first loom with a cast iron frame, the Roberts Loom. In 1842 James Bullough and William Kenworthy, made the Lancashire Loom. It is a semi automatic power loom. Although it is self-acting, it has to be stopped to recharge empty shuttles. It was the mainstay of the Lancashire cotton industry for a century, when the Northrop Loominvented in 1894 with an automatic weft replenishment function gained ascendancy.

Number of Looms in UK

Year	1803	1820	1829	1833	1857
Looms	2400	14650	55500	100000	250000

Also in 1830, Richard Roberts patented the first self-acting mule. The Stalybridge mule spinners strike was in 1824, this stimulated research into the problem of applying power to the winding stroke of the mule. The draw while spinning had been assisted by power, but the push of the wind had been done manually by the spinner, the mule could be operated by semiskilled labour. Before 1830, the spinner would operate a partially powered mule with a maximum of 400 spindles after, self-acting mules with up to 1300 spindles could be built.

The savings that could be made with this technology were considerable. A worker spinning cotton at a hand-powered spinning wheel in the 18th century would take more than 50,000 hours to spin 100 lb of cotton; by the 1790s, the same quantity could be spun in 300 hours by mule, and with a self-acting mule it could be spun by one worker in just 135 hours.

WORKING PRACTICES

The nature of work changed during industrialisation from a craft production model to a factory-centric model. It was during the years 1761 to 1850 that these changes happened. Textile factories organized workers' lives much differently from craft production. Handloom weavers worked at their own pace, with their own tools, and within their own cottages. Factories set hours of work, and the machinery within them shaped the pace of work.

Factories brought workers together within one building to work on machinery that they did not own. Factories also increased the division of labour. They narrowed the number and scope of tasks They included children and women within a common production process. As Manchester mill owner Friedrich Engels decried, the family structure itself was "turned upside down" as women's wages undercut men's, forcing men to "sit at home" and care for children while the wife worked long hours.

Factories flourished over manual craftsmanship because they had more efficient production output per worker, keeping prices down for the public, and they had much more consistent quality of product. The work-discipline was forcefully instilled upon the workforce by the factory owners, and he found that the working conditions were poor, and poverty levels were at an unprecedented high. Engels was appalled, and his research in Derby played a large role in his andMarx's book 'Das Kapital'. At times, the workers rebelled against poor wages.

The first major industrial action in Scotland was that of the Calton weavers in Glasgow, who went on strike for higher wages in the summer of 1787. In the ensuing disturbances, troops were called in to keep the peace and three of the weavers were killed. There was continued unrest. In Manchester in May 1808, 15,000 protesters gathered on St George's Fields and were fired on by dragoons, with one man dying. A strike followed, but was eventually settled by a small wage increase. In the general strike of 1842, half a million workers demanded the Charter and an end to pay cuts. Again, troops were called in to keep the peace, and the strike leaders were arrested, but some of the worker demands were met.

The early textile factories employed a large share of children, but the share declined over time. In England and Scotland in 1788, two-thirds of the workers in 143 water-powered cotton mills were described as children. Sir Robert Peel, a mill owner turned reformer, promoted the 1802 Health and Morals of Apprentices Act, which was intended to prevent pauper children from working more than 12 hours a day in mills.

Children had started in the mills at around the age of four, working as mule scavengers under the working machinery until they were eight, they progressed to working as little piecers which they did until they were 15. During this time they worked 14 to 16 hours a day, being beaten if they fell asleep. The children were sent to the mills of Derbyshire, Yorkshire and Lancashire from the workhouses in London and other towns in the south of England. A well documented example was that of Litton Mill.

Further legislation followed. By 1835, the share of the workforce under 18 years of age in cotton mills in England and Scotland had fallen to 43%. About half of workers in Manchester and Stockport cotton factories surveyed in 1818 and 1819 had begun work at under ten years of age. Most of the adult workers in cotton factories in mid-19th-century Britain were workers who had begun work as child labourers. The growth of this experienced adult factory workforce helps to account for the shift away from child labour in textile factories.

A REPRESENTATIVE EARLY SPINNING MILL 1771

Cromford Mill was an early Arkwright mill and was the model for future mills. The site at Cromford had year-round supply of warm water from the sough which drained water from nearby lead mines, together with another brook. It was a five storey mill. Starting in 1772, the mills ran day and night with two 12 hour shifts.

Fig. Gateway to Arkwright's Mill

It started with 200 workers, more than the locality could provide so Arkwright built housing for them nearby, one of the first manufacturers to do so. Most of the employees were women and children, the youngest being only 7 years old. Later, the minimum age was raised to 10 and the children were given 6 hours of education a week, so that they could do the record keeping their illiterate parents could not.

The first stage of the spinning process is carding, initially this was done by hand, but in 1775 he took out a second patent for a water-powered carding machine and this led to increased output. He was soon building further mills on this site and eventually employed 1,000 workers at Cromford. By the time of his death in 1792, he was the wealthiest untitled person in Britain. The gate to Cromford Mill was shut at precisely 6am and 6pm every day and any worker who failed to get through it not only lost a day's pay but was fined another day's pay. In 1779, Arkwright installed a cannon, loaded with grapeshot, just inside the factory gate, as a warning to would-be rioting textile workers, who had burned down another of his mills in Birkacre, Lancashire. The cannon was never used.

The mill structure is classified as a Grade I listed building, it was first classified in June 1950.

A REPRESENTATIVE MID-CENTURY SPINNING MILL 1840

Brunswick Mill, Ancoats is a cotton spinning mill in Ancoats, Manchester, Greater Manchester. It was built around 1840, part of a group of mills built along the Ashton Canal, and at that time it was one of the country's largest mill. It was built round a quadrangle, a seven storey block faced the canal. It was taken over by the Lancashire Cotton Corporation in the 1930s and passed to Courtaulds in 1964. Production finished in 1967. The Brunswick mill was built around 1840 in one phase. The main seven storey block that faces the

Ashton Canal was used for spinning. The preparation was done on the second floor and the self-acting mules with 400 spindles were arranged transversely on the floors above. The wings contained the blowing rooms, some spinning and ancillary processes like winding. The four storey range facing Bradford Road was used for warehousing and offices. The mill was built by David Bellhouse, but it is suspected that William Fairbairn was involved in the design. It is built from brick, and has slate roofs. Fireproof internal construction was now standard. Brunswick was built using cast iron columns and beams, each floor was vaulted with transverse brick arches. There was no wood in the structure. It was powered by a large double beam engine.

In 1850 the mill had some 276 carding machines, and 77,000 mule spindles, 20 drawing frames, fifty slubbing frames and eighty one roving frames.

The structure was good and it successfully converted to ring spinning in 1920- and was the first mill to adopt mains electricity as its principal source of power. The mill structure was classified as a Grade II listed building in June 1994.

EXPORT OF TECHNOLOGY

While profiting from expertise arriving from overseas (e.g. Louis Paul), Britain was very protective of home-grown technology. In particular, engineers with skills in constructing the textile mills and machinery were not permitted to emigrate — particularly to the fledgeling America.

Horse power (1780–1790)

The earliest cotton mills in the United States were horse powered. The first mill to use this method was the Beverly Cotton Manufactory, built in Beverly, Massachusetts. It was started August 18, 1788 by entrepreneur John Cabot and brothers. It was operated in joint by Moses Brown, Israel Thorndike, Joshua Fisher, Henry Higginson, and Deborah Higginson Cabot. The Salem Mercury reported that in April 1788 that the equipment for the mill was complete, consisting of a spinning jenny, a carding machine, warping machine, and other tools. That same year the mill's location was finalized and built in the rural outsets of North Beverly. The location had the presence of natural water, but it was cited the water was used for upkeep of the horses and cleaning of equipment, and not for mass-production.

Much of the internal designs of the Beverly mill were hidden due to concerns of competitors stealing designs. The beginning efforts were all researched behind closed doors, even to the point that the owners of the mill set up milling equipment on their estates to experiment with the process. There were no published articles describing exactly how their process worked in detail. Additionally, the mill's horse powered technology was quickly dwarfed by new water-powered methods.

Slater

Following the creation of the United States, an engineer who had worked as an apprentice to Arkwright's partner Jedediah Strutt evaded the ban. In 1789, Samuel Slater took his skills in designing and constructing factories to New England, and he was soon engaged in reproducing the textile mills that helped America with its own industrial revolution.

Local inventions spurred this on, and in 1793 Eli Whitney invented and patented the cotton gin, which sped up the processing of raw cotton by over 50 times.

TEXTILE MANUFACTURING BY PRE-INDUSTRIAL METHODS

Textile manufacturing is one of the oldest human activities. The oldest known textiles date back to about 5000 B.C. In order to make textiles, the first requirement is a source of fibre from which a yarn can be made, primarily by spinning. The yarn is processed by knitting or weaving to create cloth. The machine used for weaving is the loom. Cloth isfinished by what are described as wet processes to become fabric. The fabric may be dyed, printed or decorated by embroidering with coloured yarns.

The three main types of fibres are natural vegetable fibres, animal protein fibres and artificial fibres. Natural vegetable fibres include cotton, linen, jute and hemp. Animal protein fibres include wool and silk. Man-made fibres (made by industrial processes) including nylon, polyester will be used in some hobbies and hand crafts and in the developed world.

Almost all commercial textiles are produced by industrial methods. Textiles are still produced by pre-industrial processes in village communities in Asia, Africa and South America. Creating textiles using traditional manual techniques is an artisan craft practised as a hobby in Europe and North America.

YARN FORMATION

VEGETABLE FIBRES

Flax

The preparations for spinning is similar across most plant fibres, including Flax and Hemp. Flax is the fibre used to create linen. Cotton is handled differently since it uses the fruit of the plant and not the stem.

Harvesting

Flax is pulled out of the ground about a month after the initial blooming when the lower part of the plant begins to turn yellow, and when the most forward of the seeds are found in a soft state. It is pulled in handfuls and several handfuls are tied together with slip knot into a 'beet'. The string is tightened as the stalks dry. The seed heads are removed and the seeds collected, by threshing and winnowing.

Retting

Retting is the process of rotting away the inner stalk, leaving the outer fibres intact. A standing pool of warm water is needed, into which the beets are submerged. An acid is produced when retting, and it would corrode a metal container.

At 80 °F (27 °C), the retting process takes 4 or 5 days, it takes longer when colder. When the retting is complete the bundles feel soft and slimy, The process can be overdone, and the fibres rot too.

Dressing the flax

Dressing is removing the fibres from the straw and cleaning it enough to be spun. The flax is broken, scutched and hackled in this step. Breaking The process of breaking breaks up the straw into short segments. The beets are untied and fed between the beater of the breaking machine , the set of wooden blades that mesh together when the upper jaw is lowered. Scutching In order to remove some of the straw from the fibre a wooden scutching knife is scaped down the fibres while they hang vertically.

Heckling Fibre is pulled through various sized heckling combs. A Heckling comb is a bed of sharp, long-tapered, tempered, polished steel pins driven into wooden blocks at regular spacing. A good progression is from 4 pins per square inch, to 12, to 25 to 48 to 80. The first three will remove the straw, and the last two will split and polish the fibres. Some of the finer stuff that comes off in the last heckles can be carded like wool and spun. It will produce a coarser yarn than the fibres pulled through the heckles because it will still contain some straw.

Spinning

Flax can either be spun from a distaff, or from the spinner's lap. Spinners keep their fingers wet when spinning, to prevent forming fuzzy thread. Usually singles are spun with an "S" twist. After flax is spun it is washed in a pot of boiling water for a couple of hours to set the twist and reduce fuzziness. Many handspinners, will buy a roving of flax. This roving is spun in the same manner as above. The rovings may come with very long fibres (4 to 8 inches), or much shorter fibres (2 to 3 inches).

Cotton

Cotton is a soft, fluffy staple fibre that grows in a boll, or protective capsule, around the seeds of cotton plants of the genus Gossypium. The fibre is almost pure cellulose. The plant is a shrub native to tropical and subtropical regions around the world, including the Americas, Africa, and India. The greatest diversity of wild cotton species is found in Mexico, followed by Australia and Africa. Cotton was independently domesticated in the Old and New Worlds. The most favoured cottons are the ones with the longest staple as they can be spun into the finest thread. Sea Islandand Egyptian are two of these. Surat an Indian species has a short staple. Hand operated methods of processing remained the preferred way of spinning and weaving the very finest threads and fabrics into the third quarter of the nineteenth century.

Yucca

Yucca fibres were at one time widely used throughout Central America for many things. Currently they are mainly used to maketwine.Yucca leaves are harvested and then cut to a standard size. The leaves are crushed in between two large rollers producing the fibres which are bundled up and dried in the sun over trellises. The dried fibres are combined into rolags. At this point it is ready to spin. The waste, a pulpy liquid that stinks, can be used as a fertilizer.

ANIMAL PROTEIN FIBRES

Wool

Wool is a protein based fibre, being the coat of a sheep. The wool is removed by shearing.

Sheep Shearing

Shearing can be done with use of hand-shears or powered shears. Professional sheep shearers can shear a sheep in under a minute, without nicking the sheep.

The fleece is removed in one piece. Second cuts can be made but produce only short fibres, which are more difficult to spin. Primitive breeds, like the Scottish Soay sheep have to be plucked, not sheared, as the kemps are still longer than the soft fleece, (a process called rooing).

Skirting

Skirting is disposing of all wool that is unsuitable for spinning. Recovering can be attempted. It can also be done at the same time as carding.

Cleaning

The wool is cleaned. At this point the fleece is full of lanolin and often contains extraneous vegetable matter, such as sticks, twigs, burrs and straw. These may all be removed, though lanolin may be left in the wool till after the spinning, a technique known as spinning 'in the grease'. Indeed if the fabric is to be water repellent, lanolin is not removed at any stage. Washing the wool at this stage can be a tedious process. Some people wash it a small handful at a time very carefully, and then set it out to dry on a table in the sun. Others will wash the whole fleece. Lanolin is removed by soaking the fleece in very hot water. If the fleece gets agitated, it will become felt, and then spinning is impossible. Felting, when done on purpose (with needles, chemicals, or simply rubbing the fibres against each other), can be used to create garments.

Carding or combing

It is possible to spin directly from a clean fleece, but it is much easier to spin a carded fleece. Carding by hand yields a rolag, a loose woollen roll of fibres. Using a drum carder yields a bat, which is a mat of fibres in a flat, rectangular shape.

Carding mills return the fleece in a roving, which is a stretched bat; it is very long and often the thickness of a wrist.A pencil roving is a roving thinned to the width of a pencil. It can used for knitting without any spinning, or for apprentices.

Combing is another method to align the fibres parallel to the yarn, and thus is good for spinning a worsted yarn, whereas the rolag from handcards produces a woolen yarn.

Spinning

Fig. A spinning wheel used to make yarn.

Hand spinning can be done by using a spindle or the spinning wheel. Spinning turns the carded wool fibres intoyarn which can then be directly woven, knitted (flat or circular), crocheted, or by other means turned into fabric or a garment. The spinning wheel collects the yarn on a bobbin. A woollen yarn is lightly spun so it is airey, and is a good insulator and suitable for knitting, while a worsted yarn is spun tight to exclude air, and has greater strength and is suited to weaving..

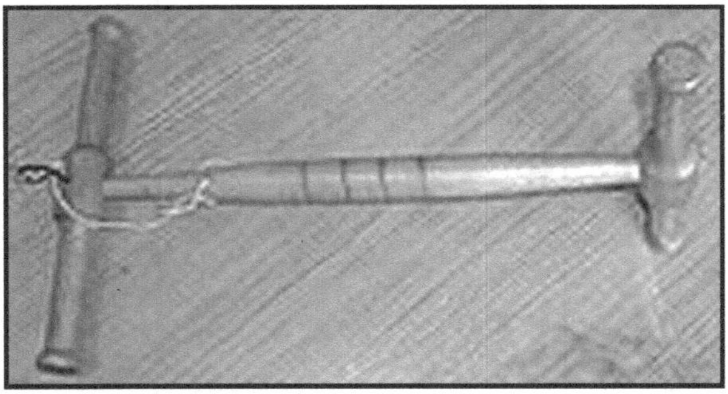

Fig. A niddy noddy ready to have a skein wound on it.

Once the bobbin is full, the hobby spinner either puts on a new bobbin, or forms a skein, or balls the yarn. A skein is a coil of yarn twisted into a loose knot. Yarn is skeined using a niddy-noddy or other type of skein -winder. Yarn is rarely balled directly after spinning, it will be stored in skein form, and transferred to a ball only if needed. Knitting from a skein, is difficult as the yarn forms knots, in this case it is best to ball. Yarn to be plied is left on the bobbin.

Fig. A lazy kate with bobbins on it in preparation for plying.

A skein is either formed by the hobby spinner, on a niddy noddy or some other type of skein winder. Traditionally niddy-noddys looked like an uppercase "i", with the bottom half rotated 90 degrees. Hobby spinning wheel manufactures also make niddy-noddys that attach onto the spinning wheel for faster skein winding.

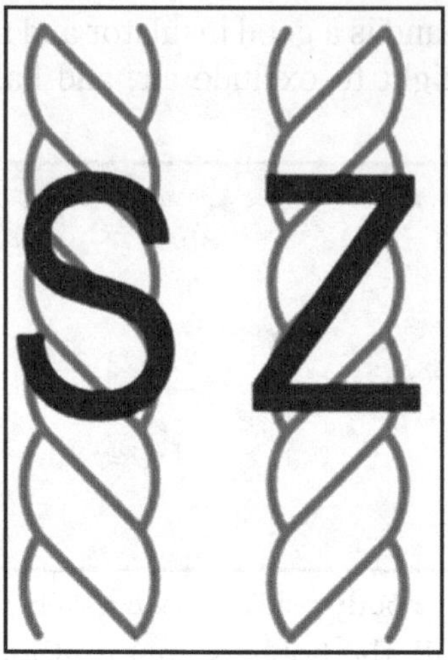

Fig. S and Z twists

Plying

Plying yarn is when one takes a strand of spun yarn (one strand is often called a single) and spins it together with other strands in order to make a thicker yarn. Regular plying consists of taking two or more singles and twisting them together, the against their twist. This can be done on a spinning wheel or on a spindle. If the yarn was spun clockwise (which is called a "Z" twist), to ply, the wheel must spin counter-clockwise (an "S" twist). This is the most common way. When plying from bobbins a device called a lazy kate is often used to hold them.

Most hobby spinners (who use spinning wheels) ply from bobbins. This is easier than plying from balls because there is less chance for the yarn to become tangled and knotted if it is simply unwound from the bobbins. So that the bobbins can unwind freely, they are put in a device called a lazy kate, or sometimes simply kate. The simplest lazy kate consists of wooden bars with a metal rod running between them. Most hold between three and four bobbins. The bobbin sits on the metal rod. Other lazy kates are built with devices that create an adjustable amount of tension, so that if the yarn is jerked, a whole bunch of yarn is not wound off, then wound up again in the opposite direction. Some spinning wheels come with a built in lazy kate.

Navajo plying consists of making large loops, similar to crocheting.A loop about 8 inches long is made on the leader the end on the leader. (A leader is the string left on the bobbin to spin off.) The three strands together are spun in the opposite direction. When a third of the loop remains, a new loop is created and the spinning continues. The process is repeated until the yarn is all plied. The advantage of this method is that only one single is needed and if the single is already dyed this technique allows it to be plied without ruining the color scheme. This technique also allows the spinner to try to match up thick and thin spots in the yarn, thus making for a smoother end product.

Washing

If the lanolin is unwanted, and has not already been washed out, this is done now. The skein is tied in six points and steeped overnight in detergent, it is rinsed and air-dried, and re-skeined.

unless the lanolin is to be left in the cloth as a water repellent. When washing a skein it works well to let the wool soak in soapy water overnight, and rinse the soap out in the morning. Dishwashing detergents are commonly used, and a special laundry detergent designed for washing wool is not required. The dishwashing detergent works and does not harm the wool. After washing, let the wool dry (air drying works best). Once it is dry, or just a bit damp, one can stretch it out a bit on a niddy-noddy. Putting the wool back on the niddy-noddy makes for a nicer looking finished skein. Before taking a skein and washing it, the skein must be tied up loosely in about six places. If the skein is not tied up, it will be very hard to unravel when done washing.

Silk

Silk fabric was first developed in ancient China, with some of the earliest examples found as early as 3500 BC.

Cultivation

Silk moths lay eggs on specially prepared paper. The eggs hatch and the caterpillars (silkworms) are fed on fresh mulberry leaves. After about 35 days and 4 moltings, the caterpillars are 10,000 times heavier than when hatched and are ready to begin spinning a cocoon. A straw frame is placed over the tray of caterpillars, and each caterpillar begins spinning a cocoon by moving its head in a pattern. Two glands produce liquid silk and force it through openings in the head called spinnerets. Liquid silk is coated insericin, a water-soluble protective gum, and solidifies on contact with the air. Within 2–3 days, the caterpillar spins about 1 mile of filament and is completely encased in a cocoon.

Harvesting

The silk farmers then kill most caterpillars by heat, leaving some to metamorphose into moths to breed the next generation of caterpillars. Harvested cocoons are then soaked in boiling water to soften the sericin holding the silk fibres together in a cocoon shape.

Throwing

The fibres are then unwound to produce a continuous thread. Since a single thread is too fine and fragile for commercial use, anywhere from three to ten strands are spun together to form a single thread of silk. Colloquially silk throwing can be used to refer to the whole process: reeling, throwing and doubling, and silk throwsters would speak of throwing as twisting or spinning.

Fig. 1843 Illustration

Silk throwing was originally a hand process relying on a turning a wheel (the gate) that twisted four threads while a helper who would be a child, ran the length of a shade, hooked the threads on stationary pins (the cross)and ran back to start the process again. The shade would be a between 23 and 32m long. The process was described in detail to Lord Shaftesbury's Royal Commission of Inquiry into the Employment of Children in 1841:

For twisting it is necessary to have what are designated shades which are buildings of at least 30 or 35 yards in length... the upper storey is generally occupied by children... or grown women as 'piecers', 'winders' and 'doublers' attending to their reels and bobbins [which is], driven by the exertions of one man... He (the boy) takes first a rod containing four bobbins of silk from the twister who stands at his gate or wheel, and having fastened the ends, runs to the 'cross' at the extreme end of the room, round which he passes the threads of each bobbin and returns to the 'gate'.

He is despatched on a second expedition of the same kind, and returns as before, he then runs up to the cross and detaches the threads and comes to the roller. Supposing the master to make twelve rolls a day, the boy necessarily runs fourteen miles, and this is barefooted.

In 1700, the Italians were the most technologically advanced throwsters in Europe and had developed two machines capable of winding the silk onto bobbins while putting a twist in the thread. They called the throwing machine, a filatoio, and called the doubler, a torcitoio.

There is an illustration of a circular hand-powered throwing machine drawn in 1487 with 32 spindles. The first evidence of an externally powered filatoio comes from the thirteenth century, and the earliest illustration from around 1500.

Filatorios and torcitoios contained parallel circular frames that revolved round each other on a central axis. The speed of the relative rotation determined the twist. Silk would only cooperate in the process if the temperature and humidity were high.

FABRIC FORMATION

Once the fibre has been turned into yarn the process of making cloth is much the same for any type of fibre, be it animal or plant.

Weaving

The earliest weaving was done without a loom.

Loom

Elements of a foot-treadle floor loom

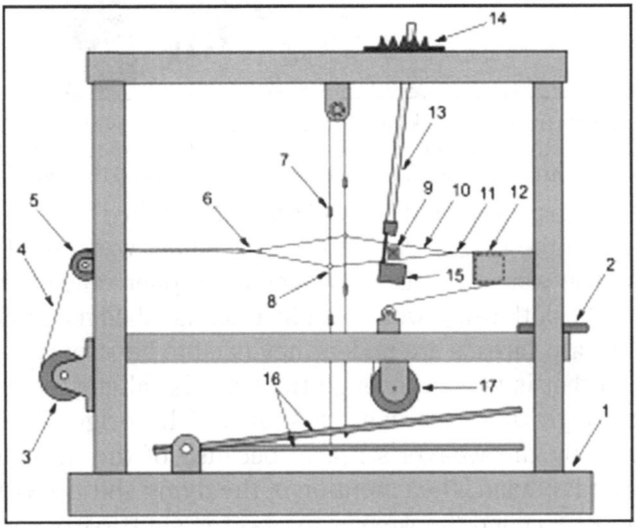

1. Wood frame
2. Seat for weaver
3. Warp beam
4. Warp threads
5. Back beam or platen
6. Rods – used to make a shed
7. Heddle bar
8. Heddle- the eye
9. Shuttle with weft yarn
10. Shed
11. Completed fabric
12. Breast beam
13. Batten with reed comb
14. Batten adjustment
15. Lathe
16. Treadles
17. Cloth roll

Fig. A picture taken from the back of a loom. The metal rods with holes that have the yarn running through them are the heddles. Further back, the metal comb with wood on the top and bottom is the reed. The shed is the gap between the two sets of yarn.

In general the supporting structure of the loom is called the frame. It provides the means of fixing the length-wise threads, called the warp, and keeping them under tension. The warp threads are wound on a roller called the warp beam, and attached to thecloth beam which will hold the finished material. Because of the tension the warp threads are under, they need to be strong.

The thread that is woven through the warp is called the weft. The weft is threaded through the warp using a shuttle. The originalhand-loom was limited in width by the weaver's reach, because of the need to throw the shuttle from hand to hand.The invention of the flying shuttle with its fly cord and picking sticks enabled the weaver to pass the shuttle from a box at either side of the loom with one hand, and across a greater width. The invention of the drop box allowed a weaver to use multiple shuttles to carry different wefts.

Alternating sets of threads are lifted by connecting them with string or wires called heddles to another bar, called the shaft (orheddle bar or heald). Heddles, shafts and the couper (lever to lift the assembly) are called the harness — the harness provides for mechanical operation using foot- or hand-operated treadles. After passing a weft thread through the warp, a reed comb is used to beat (compact) the woven weft.

To prepare to weave, the warp must be made. By hand this is done with the help of a warping board. The length the warp is made is about a quarter to half yard more than the amount of cloth needed. Warping boards come in a variety of shapes, from the two nearest door handles to a board with pegs on it, or a device called a warping mill that looks similar to a swift.

Warping the loom, mean threading each end through an eye in a heddle, and then sleying it through the reed. The warp is set (verb) at Xends per inch. It then has a sett (noun) of X ends per inch. The weft is measured in picks per inch.

Knitting

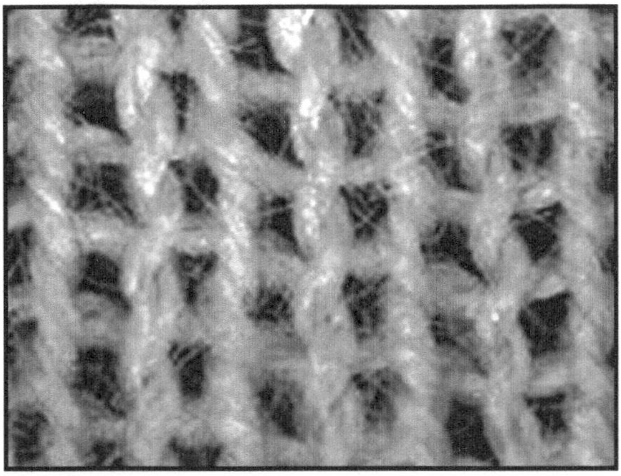

Fig. The front side of a plainly knitted object

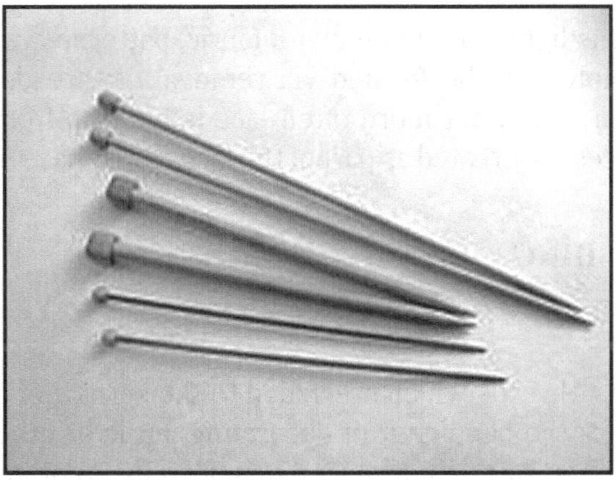

Fig. Knitting needles

Hand knitting can either be done "flat" or "in the round". Flat knitting is done on a set of single pointknitting needles, and the knitter goes back and forth, adding rows. In Circular knitting, or "knitting in the round", the knitter knits around a circle, creating a tube. This can be done with a set of four double pointed needles or a single circular needle.

A knitted object will unravel easily if the top has not been secured. Knitted objects also stretch easily in all directions, whereas woven fabric only stretches on the bias.

Crocheting

Crocheting differs largely from knitting in that there is only one loop, not the multitude as knitting has. Also, instead of knitting needles, a crochet hook is used.

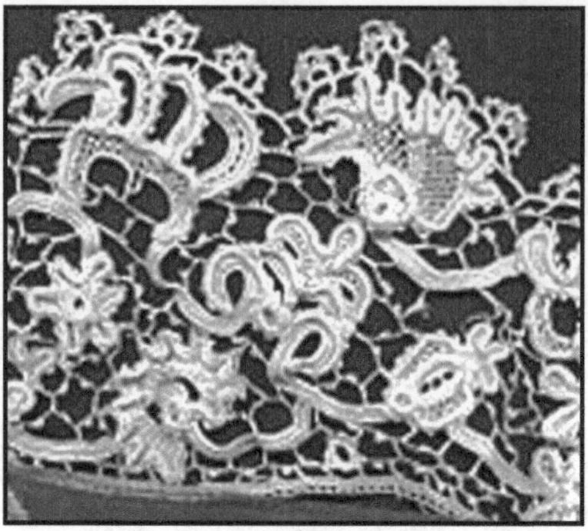

Fig. Irish crocheted lace

Lace making

A lace fabric is lightweight openwork fabric, patterned, with open holes in the work. The holes can be formed via removal of threads or cloth from a previously woven fabric, but more often lace is built up from a single thread and the open spaces are created as part of the lace fabric. Lace may be crocheted, or knitted.

TEXTILE FINISHING

Embroidery

Embroidery – threads which are added to the surface of a finished textile.

Embroidery is the handicraft of decorating fabric or other materials with needle and thread or yarn. Embroidery may also incorporate other materials such as metal strips, pearls, beads, quills, and sequins. Embroidery is most often used on caps, hats, coats, blankets, dress shirts, denim, stockings, and golf shirts. Embroidery is available with a wide variety of thread or yarn color.

2

Cotton Mill

A cotton mill is a factory housing powered spinning or weaving machinery for the production of yarn or cloth from cotton, an important product during the Industrial Revolution when the early mills were important in the development of the factory system.

Although some were driven by animal power, most early mills were built in rural locations near to fast-flowing rivers and streams and hadwater wheels to power them. The development of viable rotative steam engines by Boulton and Watt led from 1781 to the growth of larger, steam-powered mills and allowed them to be concentrated in urban mill towns, most notably Manchester, which with neighbouringSalford had more than 50 mills by 1802.

The mechanisation of the spinning process in the early factories was instrumental in the growth of the machine tool industry, enabling the construction of larger cotton mills. Limited companies were developed to construct mills, and the trading floors of the cotton exchange in Manchester, created a vast commercial city. Mills generated employment, drawing workers from largely rural areas and expanding urban populations. They provided incomes for girls and women. Child labour was used in the mills, and the factory system led to organised labour. Poor conditions became the subject of exposés, and in England, the Factory Acts were written to regulate them.

The cotton mill, originally a Lancashire phenomenon, was copied in New England and later in the southern states of America. In the 20th century, North West England lost its supremacy to the United States, then to India and subsequently to China.

HISTORY

In the mid-16th century Manchester was an important manufacturing centre for woollens and linen and market for textiles made elsewhere. The fustian district of Lancashire, from Blackburn to Bolton, west to Wigan and Leigh and south towards Manchester, used flax and raw cotton imported along the Mersey and Irwell Navigation.

Key inventions

Fig. Lancashire cotton mill, 1914

During the Industrial Revolution cotton manufacture changed from a domestic to a mechanised industry, made possible by inventions and advances in technology. The weaving process was the first to be mechanised by the invention of John Kay's flying shuttle in 1733. The manually-operated spinning jenny was developed by James Hargreaves in about 1764 speeded up the spinning process. The roller spinning principle of Paul and Bourne became the basis of Richard Arkwright's spinning frame and water frame, patented in 1769. The principles of the spinning jenny and water frame were combined by Samuel Crompton in his spinning mule of 1779, but water power was not applied to it until 1792. Many mills were built after Arkwright's patent expired in 1783 and by 1788, there were about 210 mills in Great Britain. The development of cotton mills was linked to the development of the machinery they contained. By 1774, 30,000 people in Manchester were employed using the domestic system in cotton manufacture. Handloom weaving lingered into the mid-19th century but cotton spinning in mills relying on water power and subsequently steam power using fuel from the Lancashire Coalfield began to develop before 1800.

First mills (1740–1815)

The Paul-Wyatt mills

The first cotton mills were established in the 1740s to house roller spinning machinery invented by Lewis Paul and John Wyatt. The machines were the first to spin cotton mechanically "without the intervention of human fingers". They were driven by a single non-human power source which allowed the use of larger machinery and made it possible to concentrate production into organised factories. Four mills were set up to house Paul and Wyatt's machinery in the decade following its patent in 1738: the short-lived, animal-powered Upper Priory Cotton Mill in Birmingham in 1741; Marvel's Mill in Northampton operated from 1742 until 1764 and was the first to be powered by a water wheel; Pinsley Mill in Leominster probably opened in 1744 and operated until it burned down in 1754; and a second mill in Birmingham set up by Samuel Touchet in 1744, about which little is known, but which was sufficiently successful for Touchet later to seek the lease on the mill in Northampton. The Paul-Wyatt mills spun cotton for several decades but were not very profitable, becoming the ancestors of the cotton mills that followed.

Arkwright-type mills

Richard Arkwright obtained a patent for his water frame spinning machinery in 1769. Although its technology was similar to that ofLewis Paul, John Wyatt, James Hargreaves and Thomas Highs, Arkwright's powers of organisation, business acumen and ambition established the cotton mill as a successful business model and revolutionary example of the factory system. Arkwright's first mill – powered by horses in Nottingham in 1768 – was similar to Paul and Wyatt's first Birmingham mill although by 1772 it had expanded to four storeys and employed 300 workers. In 1771, while the Nottingham mill was at an experimental stage, Arkwright and his partners started work on Cromford Mill in Derbyshire, which "was to prove a major turning point in the history of the factory system". It resembled the Paul-Wyatt water-powered mill at Northampton in many respects, but was built on a different scale, influenced by John Lombe's Old Silk Mill in Derby and Matthew Boulton's Soho Manufactory in Birmingham. Constructed as a five-storey masonry box; high, long and narrow, with ranges of windows along each side and large relatively unbroken internal spaces, it provided the basic architectural prototype that was followed by cotton mills and English industrial architecture through to the end of the 19th century.

Arkwright recruited large, highly disciplined workforces for his mills, managed credit and supplies and cultivated mass consumer markets for his products. By 1782 his annual profits exceeded £40,000, and by 1784 he had opened 10 more mills. He licensed his technology to other entrepreneurs and in 1782 boasted that his machinery was being used by "numbers of adventurers residing in the different counties of Derby, Leicester, Nottingham, Worcester, Stafford, York, Hertford and Lancashire" and by 1788 there were 143 Arkwright-type mills nationwide.

The early mills were of light construction, narrow – about 9 feet (2.7 m) wide – and low in height, with ceiling heights of only 6–8 ft. The mills were powered by water wheels and lit by daylight. Mills were made by millwrights, builders and iron founders. By the end of the 18th century there were about 900 cotton mills in Britain, of which approximately 300 were large Arkwright-type factories employing 300 to 400 workers, the rest, smaller mills using jennies or mules, were hand- or horse-driven and employed as few as 10 workers.

Early steam mills

Before 1780, only water power was available to drive large mills, but they were dependent on a constant flow of water and built in rural locations, causing problems of labour supply, transportation of materials and access to urban merchants for large mill-owners. Steam engines had been used to pump water since the invention of the atmospheric engine by Thomas Newcomen in 1712 and, starting with the engine installed at Arkwright's Haarlem Mill in Wirksworth, Derbyshire in 1780, were used to supplement the supply of water to the water wheels of cotton mills.

In 1781 James Watt registered a patent for the first rotative steam engine designed to "give motion to the wheels of mills or other machines". Concerns remained over the smoothness of the power supplied by a steam engine to cotton mills, where the regularity of the yarn produced was dependent on the regularity of the power supply, and it was not until 1785 at Papplewick, in Robinson's Mill near Nottingham that a steam engine was successfully used to drive a cotton mill directly. Boulton and Watt's engines enabled mills to be built in urban contexts and transformed the economy of Manchester, whose importance had previously been as a centre of pre-industrial spinning and weaving based on the domestic system. Manchester had no cotton mills until the opening of Arkwright'sShudehill Mill in 1783 and in 1789 Peter Drinkwater opened the Piccadilly Mill – the town's first mill to be directly powered by steam – and by 1800 Manchester had 42 mills, having eclipsed all rival textile centres to become the heart of the cotton manufacturing trade.

Water continued to be used to drive rural mills but mills, driven by steam, were built in towns alongside streams or canals to provide water for the engine. Murrays' Millsalongside the Rochdale Canal, in Ancoats were powered by 40 hp Boulton and Watt beam engines. b Some were built as room and power mills, which let space to entrepreneurs. The mills, often 'L' or U-shaped, were narrow and multi-storeyed. The engine house, warehousing and the office were inside the mill, although stair towers were external. Windows were square and smaller than in later mills. The walls were of unadorned rough brick. Construction was sometimes to fireproof designs. The mills are distinguished from warehouses in that warehouses had taking-in doors on each storey with an external hoist beam. Only the larger mills have survived.

Mills of this period were from 25 to 68 m long and 11.5 m to 14 m wide. They could be eight storeys high and had basements and attics. Floor height varied from 3.3 to 2.75 m on the upper storeys.

Boilers were of the wagon type; chimneys were square or rectangular, attached to the mill, and in some cases part of the stair column. The steam engines were typically low-pressure single-cylinder condensing beam engines. The average power in 1835 was 48 hp. Power was transmitted by a main vertical shaft with bevel gears to the horizontal shafts. The later mills had gas lighting using gas produced on site. The mules with 250–350 spindles were placed transversely to get as much light as possible.

Early weaving mills

The development of mills to mechanise the weaving process was more gradual partly because of the success of John Kay's 1733 invention of the flying shuttle, which increased the productivity of domestic hand loom weavers. Kay took out a patent for the application of water power to a Dutch loom in 1745 and opened a weaving factory in Keighley in 1750, but nothing is known of its success. A further attempt to mechanise the weaving process took place at Garrett Hall in Manchester in 1750 but was unsuccessful in enabling one worker to operate more than a single loom. The first feasible power loom was patented by Edmund Cartwright in 1785, although it was initially a primitive device it established the basic principle that would be used in powered weaving until the 20th century. In 1788 Cartwright opened Revolution Mill in Doncaster which was powered by a Boulton and Watt steam engine and had 108 power looms on three floors as well as spinning machinery, but it was not a commercial success and closed in 1790. A second mill using Cartwright's machinery, opened in Manchester in 1790 but was burned to the ground by hand loom weavers within two years. By 1803 there were only 2,400 power looms operating in Britain.

Early American mills

In the United States, the early horse-powered Beverly Cotton Manufactory was designed by Thomas Somers, who started construction and testing of the facility in 1787, finishing the factory's equipment in 1788. Experience from this factory led Moses Brown of Providence to request the assistance of a person skilled in water-powered spinning. Samuel Slater, an immigrant and trained textile worker from England, accepted Brown's proposal, and assisted with the design and construction of Slater Mill, built in 1790 on the Blackstone River in Pawtucket, Rhode Island. Slater evaded restrictions on emigration put in place to allow England to maintain its monopoly on cotton mills. Slater Mill resembled the Beverly Cotton Manufactory and a mill in Derbyshire in which he had worked.

REMODELLING AND EXPANSION (1815–1855)

United Kingdom

From 1825 the steam engine was able to power larger machines constructed from iron using improved machine tools. Mills from 1825 to 1865 were generally constructed with wooden beamed floors and lath and plaster ceilings. William Fairbairn experimented with cast iron beams and concrete floors. Mills were of red brick or sometimes local stone with a greater attention to decoration and the main gate was often highlighted with stone decoration.

The stair columns were exterior to the main floors. During this period the mules got wider and the width of the bays increased. Specialised mill architects appeared.

Mills of this period were tall, narrow, and wide. They were commonly built with one or two wings to form an 'L' or 'U' shape. Brunswick Mill was a 28-bay mill, 6 storeys of 16 m by 92 m. Each self-acting spinning mule had 500 spindles. Single-storey north light weaving sheds were sometimes added to the mills. The looms caused vibrations that damaged the structure of multi-storey buildings, and specialised weaving mills became common. They were single-storey sheds with an engine house and offices, and preparation and warehousing in a two-storey ancillary building.

Large mills remained the exception during this period. In 1833 the largest mill was that of McConnel and Company in Ancoats, Manchester with 1,545 workers, but in 1841 there were still only 25 mills in Lancashire with 1,000 workers or more, and the number of workers in the average mill was 193.

The Lancashire boiler was patented in 1844, and the economiser in 1845. This can been seen as a square brick structure between the boiler house and the chimney. The engines were double compound upright beam engines of the type patented by McNaught in 1845. Each room in the mill would have line shafts suitable for the type of frame, connected by belt drives or gearing.

In 1860, there were 2650 cotton mills in the Lancashire region, employing 440,000 people. The workers, 90 per cent of whom were adults and 56 per cent females, were paid a total of 11.5 million per annum. The mills used 300,000 hp of power, of which 18,500 was generated by waterpower. The mills had 30,387,467 spindles and 350,000 power looms. The industry imported 1,390,938,752 lb of raw cotton a year. It exported 2,776,218,427 yards of cotton cloth and 197,343,655 pounds (89,513,576 kg) of twist and yarn. The total value of its exports was 32,012,380.

1860 saw the end of this period of rapid growth. The Cotton Famine of 1861–1865 was a period when American long staple cotton became unavailable due to an American Civil War. After the war, the economics of the industry had changed, and a new larger mill was required.

United States

In 1814 the Boston Manufacturing Company of New England established a "fully integrated" mill on the Charles River at Waltham, Massachusetts. Despite the ban on exporting technology from the UK, one of its proprietors, Francis Cabot Lowell, had travelled to Manchester to study the mill system and memorised some of its details. In the same year,Paul Moody built the first successful power loom in the US. Moody used a system of overhead pulleys and leather belting, rather than bevel gearing, to power his machines. The group devised the Waltham System of working, which was duplicated at Lowell,

Massachusetts and several other new cities throughout the state. Mill girls, some as young as ten, were paid less than men, but received a fixed wage for their 73 hour week. They lived in company-owned boarding houses, and attended churches supported by the companies.

In the 1840s George Henry Corliss of Providence, Rhode Island improved the reliability of stationary steam engines. He replaced slide valves with valves that used cams. TheseCorliss valves were more efficient and more reliable than their predecessors. Initially, steam engines pumped water into a nearby reservoir that powered the water wheel, but were later used as the mill's primary power source. The Corliss valve was adopted in the UK, where in 1868 more than 60 mill engines were fitted with them.

India

The large steam-powered Bowreah Cotton Mills opened at Fort Gloster near Calcutta by British interests in the 1820s, using British women to impart machine-spinning skills to the local workforce. They closed down in 1837 but reopened with Dwarkanath Tagore as a major shareholder, and by 1840 lay at the centre of a major industrial complex powered by five steam engines, that included a twist mill, foundry and a rum distillery.

GOLDEN AGE (1855–1898)

United Kingdom

Just before 1870, a mill was built by a joint-stock spinning company and this financial structure led to a new wave of mill construction. The phrase Oldham Limiteds describes these companies. Family-run firms continued to build, but grouped into associations such as the Fine Spinners' and Doublers' Association. Joseph Stott of Oldham perfected a method of fireproof floor construction using steel beams supporting brick vaults that in turn supported concrete floors that would support heavier equipment. Ring frames replaced mule frames; they were heavier and larger and were placed transversely, the floors became larger (up to 130 feet (40 m) wide) and higher to provide light. The bay size in a mill was defined by the positioning of machines. In an 1870 mill the bay was typically 10 feet 6 inches (3.20 m), and the brick vaults 5 feet 3 inches (1.60 m) though there were variations.

Engines were run at higher pressures and from 1875, powered horizontal shafts on each floor by means of ropes. This was a prominent change as a rope race had to be built running the height of the mill. The engine needed more space and the engine house, boiler house and economiser were external to the main mill. Mills continued to get bigger, and were sometimes paired; two mills being driven by one engine. Another change was the trend of having carding on one floor. To achieve this, the ground floor was extended outwards behind the mill often a full mill width. In a single mill, the crosswall divided the blowing room from the rest, as it was here that there was greatest risk of fire.

Mills became wider, Houldsworth Mill, Reddish (1865) was 35 m wide and accommodated 1200 spindle mules. It was of four storeys and had sixteen bays on each side of a central engine house; a double mill. The central block provided offices and warehousing. A mill had a range of ancillary buildings. Stair columns often extending above the mill and housed a water tank for the sprinkler system. The floors were higher allowing for taller windows. Accrington brick was used from 1890, decorated with yellow sandstone with moulded brick and terracotta features. Etched and stained glass was used in the offices. Mills were designed by specialist architects and architectural quality became a major consideration.

The power needed and provided to drive these mills was increasing. Beam engines were installed until the 1870s when horizontal engines took over. Abbey Mill Oldham (1876) needed 700 hp, Nile Mill (1896) needed 2500 hp. By the 1890, boilers produced 160 psi, and the triple expansion horizontals became standard. Chimneys were octagonal.

United States

Following the American Civil War, mills grew larger. They were built in the southern states of South Carolina, Alabama, and Mississippi where cheap labour and plentiful water power made operations profitable. Cotton could be processed into fabric where it grew, saving transportation costs. The mills were usually combination mills, (spinning and weaving) that were water powered and used a slow burn design technique. They used a belt and pulley drive system, and heavier ring frames rather than mules. At this point they only spun and wove coarse counts. The mills were mainly in open country and mill towns were formed to support them. New England mills found it increasingly difficult to compete, and as in Lancashire, went into gradual decline until bankrupted during the Great Depression. Cotton mills and their owners dominated the economy and politics of thePiedmont well into the 20th century.

Asia

The modern Indian mechanised textile industry was born in 1854, when a steam-powered mill was opened in Bombay by Cowasjee N. Davar. More followed: there were 10 by 1865 and 47 by 1875. By 1880 there were 58 mills in India employing 40,000 workers, with over 80% of them in the cities of Bombay and Ahmedabad. From the 1870s India's own markets for finished yarn and cloth ceased to be dominated by imports from Lancashire, and during the 1870s and 1880s the Bombay cotton industry began to replace exports of yarn from Britain to China.

EDWARDIAN MILLS (1898–1914)

The cotton industry was subject to cycles of boom and slump, which caused waves of mill building. There was an optimism that dictated that slumps had to be endured and then there would be a period of even greater prosperity. The limited companies took control of spinning, while the room and power system was the norm for the weaving sheds. One point of view in the 1880s was that vertically integrating the weaving sheds into new mills would reduce costs and lead to greater profits. This route had been followed in New England, where it was successful, but not in Lancashire. The industry peaked in 1907. There was a severe slump in 1908, which endured until 1918, but the years 1919 and 1920 were more profitable than the peak year of 1907 had been.

United Kingdom

Production peaked in 1912. The war of 1914–1918 put the Lancashire industry into reverse. The British government, starved of raw cotton, established mills in south Asia exporting the spinning technology - which was copied, and became a low-labour cost competitor. In Germany, Flanders and Brazil, mills were built to the designs of the Oldham architects. The only new mills were very large to benefit from the economies of scale. Older mills were re-equipped with rings, and machines were powered by individual electric motors.

Mills of this period were large, their decoration was lavish reflecting Edwardian taste and prosperity. Most mills were built for mules. Kent Mill Chadderton (1908) was a five storey, 11 bay mill, 84.6m x 43.9m. It had 90,000 spindles. Ring frames were smaller and heavier than mules so the mills were narrower with fewer storeys. Pear Mill Bredbury (1912) was planned to be a 210,000 spindle double mill. Only the first mill was completed, it had 137,000 spindles. They had more stair columns than earlier mills, it had dust flues often built into the rope race. There were two or three windows per bay. Decoration was often in terracotta and the mill name displayed in white brick on the stir tower or chimney. Stott and Sons employed Byzantine styling in Broadstone Mill, Reddish. Specialist architects built new mills and then created extensions. The last steam-powered mill, Elk Mill, was built by Arthur Turner

Mules were built with 1300 spindles, but were gradually replaced by rings.

The increasingly powerful engines required more boilers with economisers and superheaters. Mills needed reservoirs to supply the boilers and condense the steam. The chimneys were round and taller. Three types of engines were used: triple expansion horizontal cross compound engines, Inverted marine type compounds which were more compact, and Manhattans with vertical and horizontal cylinders such as the 3500 hp engine at New Pear Mill. Rope drives were used exclusively. Electricity was gradually introduced firstly on group drives driving a shaft (Little Mill, 1908), and then later on individual machines.

United States

Mills constructed in South Carolina increased in size. At Rutledge Ford the Saluda River was dammed and a power plant constructed. It was completed in 1904 before the construction of a state-of-the-art textile mill in 1906. That power plant provided for 4,800 horse power. The mill contained 30,000 spindles. By 1916 a new mill was constructed, containing 70,200 spindles and 1,300 looms. The town was named Ware Shoals. Between 1904 and 1916, the population of Ware Shoals grew from 2 men employed to maintain the newly constructed power plant, to 2,000. By the 1960s the mill employed 5,000 people. It closed in 1985.

CONSOLIDATION (1918–50)

Though business revived in 1919, a shortage of building materials restricted the building of new mills, and activity was financial with the mills seeking recapitalisation. There is no clear concession on the reason for the final decline. Some say that the cotton men concentrated on making easy money ignoring the possibility of foreign competition best countered by larger mills by re-equipping the mills with more modern ring frames. Daniels and Jewkes argued the fundamental cause of the depression was a change in demand for cotton goods. J. M. Keynes suggested that there was over capacity, and the industry should be reorganised into larger units that would scrap the excess capacity.

The Lancashire Cotton Corporation was a company set up by the Bank of England in 1929, to rescue the Lancashire spinning industry by means of consolidation. In merged 105 companies, ending up in 1950 with 53 operating mills. These were the later larger mills. It was bought up by Courtaulds in August 1964.

The later mills were on the fringe of the spinning area in Wigan and Stockport, Availability of labour was cited as a reason.The last mills were completed in 1926, these were Holden Mill(Astley Bridge Mill) and Elk Mill.

In 1929, for the first time there were more spindles in the USA than in The UK. In 1972, India had greater spindleage than the USA, and it was in turn surpassed by China in 1977.

COTTON MILLS IN THE LATE 20TH CENTURY (1950–2000)

Decline of spinning in England

Though there was a slight revival after 1945, mills closed. The most efficient mills had abandoned their steam engines, and were working the frames with individual electric motors. Broadstone Mills Stockport, was built as a double mill with 265,000 mule spindles, but by 1959 it was running 37,500 mule spindles and 70,000 ring spindles. It closed in 1959 taking advantage of the Cotton Industry Act 1959 and was then used by the John Myers mail order company.

One mill was later demolished leaving the other to be used as a Shopping Outlet Centre and Craft Village. The reduction of capacity led to a legacy of redundant mills, which were readily reused for other industrial purposes.

Ring spinning technology had successfully replaced the spinning mule, with mills having been converted mules to rings. However in the 1970s, the depleted industry was challenged by a new technology open-end or break spinning. In 1978 Carrington Viyella opened a factory to do open-end spinning in Atherton. This was the first new textile production facility in Lancashire since 1929. Immediately Pear Mill, Stockport and Alder Mill, Leigh were closed. These were both Edwardian mills designed by Stott and Sons. The mill built in 1978 was built on the Howe Bridge mills site and was named Unit One. It was not an open end mill but a combed cotton ring mill.

MODERN COTTON MILLS

Modern spinning mills are mainly built around open end spinning techniques using rotors or ring spinning techniques using spindles. In 2009 there were 202,979,000 ring spinning spindles installed worldwide, with 82% of these being in Asia or Oceania and 44% being within China. In the same year there were 7,975,000 open end spinning rotors installed, with 44% of these being within Asia or Oceania and 29% within Eastern Europe. The average age of installed rotors is much lower than that of spindles and as rotors are between 7 and 10 times more productive they are responsible for 20% of the cotton spun worldwide.

Modern cotton mills are increasingly automated. One large mill in Virginia in the United States employs 140 workers in 2013 to produce an output that would have required more than 2,000 workers in 1980.

LOCATIONS

Cotton mills were not confined to Lancashire but were built in northeast Cheshire, Derbyshire, Nottingham, the West Riding of Yorkshire, Bristol, Durham and the west of Scotland. The availability of streams or rivers to provide power determined the location of the early mills some of which were in isolated areas. In Lancashire they were built on the rivers and streams descending from the Pennines and Rossendale moorland. In some places quite small streams powered a string of small mills such as in the Cheesden Valley between Ramsbottom and Heywood. where 14 mills and their associated leats and ponds were concentrated along a four-mile stretch of the brook. Mills were built around Rochdale and Littleborough. North of Bury, ten mills occupied a mile long stretch of a stream in the Shuttleworth Valley. Other mills were built north of the River Ribble and a cluster of five mills in Caton near the port at Lancaster, one of which belonged to Samuel Greg who built Quarry Bank Mill at Styal in Cheshire. Not all water-powered mills were in rural areas, after 1780 mills were built in Blackburn and Burnley. In Scotland, four cotton mills were built in Rothsay on the Isle of Bute using labour that had experience of the linen industry. By 1800 there were two water-powered mills atGatehouse of Fleet employing 200 children and 100 adults. Robert Owen who had worked for Peter Drinkwater in Manchester, developed the mills at New Lanark built by his father-in-law, David Dale under licence from Arkwright.

Cotton Mills in 1860			
	Lancashire	**Cheshire**	**Derbyshire**
Mills	1920	200	25
Workers	310000	38000	12000

ARCHITECTURE

Fireproof construction

Cotton mills were huge fire risks, cotton fibres in the air could form an explosive mixture in their gas-lit interiors. The first mills using fireproof construction were built in Shropshire and Derbyshire in the 1790s and Philips & Lee's mill built in Salford in 1801-2. Fireproofing took the form of cast iron columns and beams from which sprang jack arches that were infilled with ash or sand and covered with stone flags or floorboards. In some mills timber was also eliminated from the roof structure which was supported by cast orwrought iron trusses. Until the properties of cast iron were properly understood some mills constructed using the early technology collapsed. In Manchester extensive testing of cast iron as a structural material was carried out by Eaton Hodgkinson and William Fairbairn in the early 1820s. Fireproof construction was expensive and timber, sometimes clad in plaster or metal continued to be used throughout the 19th century. Rolled steel beams and reinforced concrete flooring was introduced in a limited way in the 1880s but not widely adopted in Lancashire mills until the 20th century.

Other factors

Cotton is sensitive to temperature and humidity. Heating systems used wrought iron pipes suspended at a height of 7 feet (2.1 m) to carry steam under pressure. In summer the system was barely used but in winter the boilers would be fired up two hours before the shift started to warm up the mill. As heat was applied the humidity dropped and system of humidifiers, either atomisers which played an air jet against a jet of water or ones that injected a stream/air mixture into the room.

Early fire fighting systems used sprinklers supplied by water captured on flat roofs in shallow tanks. Later mills had a water tank at the top of the stair tower. Water for the sprinklers had to be protected from freezing and evaporation. Water pressure needed to be above 7 psi, and the header tank at least 15 feet (4.6 m) above the highest sprinkler. The provision of light, water tanks and heating system defined the structure and shape of the mill.

MACHINERY

POWER

The earliest cotton mills were driven by water, so needed to be situated on fast flowing streams. From about 1820, the stationary steam engine became the normal form of power for a cotton mill, water was still needed to produce the steam and to condense it, to maintain the humidity, for many of the finishing processes and for firefighting. Water was extracted from rivers and canals, then later mills requiring ever more water, built and maintained their own reservoirs.

In 1781 James Watt marketed a rotary-motion steam engine that could be adapted to drive all sorts of machinery, Richard Arkwright pioneered its use in his cotton mills. Possibly the first steam engine to be used in a cotton mill was a Newcomen engine which was used at Shudehill Mill in 1783 to raise water between storage ponds so it could drive a water wheel. By 1795 most similar engines around Manchester had been replaced by Boulton and Watt or Bateman and Sherratt engines.

Electricity was introduced in 1877. Steam engine drove generators to provide electric lighting. By the 1890s this was common. Electricity was used to drive the mills machinery by 1906. It was generated in the engine house, and one group-drive electric motor was placed on each floor to drive the shafts. Generators were placed exterior to the mill as it was thought that they were a fire risk. Mains driven mills started about 1907. Later mills used individual electric motors to power the machinry.

TRANSMISSION

Fig. A weaving shed, showing how all the looms were powered from overhead shafts

The early mills had a vertical shaft to take the power from the flywheel. On each floor horizontal shafts engaged with the main shaft using bevel gearing.

American mills used thick leather bands instead of shafts. A new approach was to use thick cotton ropes. A rope drum was attached to the flywheel with a channel cut for each rope. The profile was such to give maximum adherence.

Spinning

A spinning mill opened raw cotton bales and cleaned the cotton in the blowing room. The cotton staples are carded into lap and straightened and drawn into roving which is spun using either a mule or ring frame. The yarn can be doubled and processed into thread, or prepared for weaving.

Minerva Mill, Ashton-under-Lyne was designed by P. S. Stott and equipped by John Hetherington and Son, it produced 40's twists and 65 wefts. It was typical of a mill of the 1890s.

Number	Machines	Ratio
2	Vertical Openers and scutchers	1 : 43,434
4	Intermediate Scutchers	1 :21,717
93	Carding Engines	1 :934
63	finishing deliveries of drawing	1 :1,380
792	Slubbing spindles	1 :109.7
1716	Intemediate spindles	1 :50.6
6680	Roving spindles	1 :13
86,868	Mule spindles	1 :1

Self acting mule frame (Roberts 1830) was an improvement on Crompton's Mule (1779) which derived from earlier inventions. Mules were used in the 19th century mills for the finest counts, these needed skilled workers to operate them.

Ring frame (1929) developed out of the Throstle frame (19th century) an improvement on the Arkwright's Water frame. Originally rings were only suitable for coarse counts, they were lower and heavier than mules so needed stronger floors but lower rooms. Over time, rings became suitable for finer counts and because of cheaper labour costs they replaced mules. By 1950 all mills had converted to the Ring frame. The Hawk mill in Shaw near Oldham was still operating cotton mule frames in 1964/65.

Weaving

A weaving mill needed yarn suitable for the warp and the weft. The warp had to delivered on the beam, or was wound on the beam from cheeses by a beamer. To obtain the extra strength needed, the yarn was sized on a sizer. The weft was wound onto the pirns for the shuttle on a pirner. These preparatory processes completed the yarn was woven on a loom. One weaver would operate 4 or six looms. A self-acting loom would stop when any thread broke, and the thread had to be retied or pieced. The process required greater levels of light than spinning, and weaving sheds would often be single storey, with overhead north facing lights. Placing a loom onto the ground also reduced the problems caused by the vibrations of operation.

The Cartwright's powerloom (1785) was made reliable by Robert's cast iron power loom (1822) and became perfected by the Kenworthy and Bullough Lancashire Loom (1854). The Northrop or Draper Loom (1895) replaced these older designs.

Number of power looms in Britain 1803–57, and 1926							
Year	1803	1820	1829	1833	1857	-	1926
No. of Power Looms in UK	2,400	14,650	55,500	100,000	250,000		767,500

Comparison of output of 24 yard shirtings 1823–33				
Year	1823	1823	1826	1833
	good hand loom weaver	power weaver	power weaver	power weaver
	Aged 25	Aged 15	Aged 15	Aged 15 with 12yr old helper
Looms	1	2	2	4
Pieces woven per week	2	7	12	18

Labour conditions

The mills were notable in employing women, giving them an independent income. In Lancashire and Piedmont, South Carolina child labour is well documented.

Child labour in the United Kingdom

The Lancashire and Derbyshire mills needed a pool of cheap labour. Pauper children were boys and girls between the ages of 7 and 21, who were dependent on the Poor Law Guardians. Mill owners made contracts with the guardians in London and the southern counties to supply them paupers, in batches of 50 or more, to be apprenticed. Living condition were poor in 'Prentice Houses', and the children who were paid 2d a day worked 15 hour shifts, hot bedding with children on the other shift.

Robert Owen was a millowner in New Lanark. He never employed children under the age of ten, and opposed physical punishment in schools and factories. He lobbied for parliamentary action, resulting in The Health and Morals of Apprentices Act 1802.

- Hours of work limited to twelve a day
- Boys and girls to sleep in separate dormitories with no more than two to each bed
- Compulsory education to be provided in the arts of reading writing and arithmetic

- Each apprentice to be provided with two suits of clothes
- On Sunday children to be instructed in Christian worship
- Sanitation to be improved

Regulation was ineffective until the mills were subject to inspection in 1833 with the establishment of a factory inspectorate consisting of four factory inspectors who had the power of magistrates. This did not reduce the number of children, half-timers worked mornings in the mill and spend the afternoon in the school room. While the number of children working in spinning as tenters did decline, more were employed in weaving because weavers were expected to tenter extra looms.

Percentage of children below 13 in cotton factories 1835–78										
Year	1835	1838	1847	1850	1856	1862	1867	1870	1874	1878
amount	13.2	4.7	5.8	4.6	6.5	8.8	10.4	9.6	14.0	12.8

From the Factory Act of 1844, until 1878 records do not distinguish between full-time and half-times. In 1851 a sizeable number of children were working the mills. For Example, In Glossop, there were 931 children (out of 3562) between 5 and thirteen working in cotton mills. In one mill in 1859, 50.2% of the workforce were women, 24.2% were girls, 19% were men and 6.6% were boys.

Child labor (sic) in the United States

Fig. A little spinner in the Mollahan Mills,Newberry, South Carolina. She was tending her 'sides' like a veteran, but after I took the photo, the overseer came up and said in an apologetic tone that was pathetic, 'She just happened in.' Then a moment later he repeated the information. The mills appear to be full of youngsters that 'just happened in,' or 'are helping sister.' December 3, 1908. Witness Sara R. Hine. Location: Newberry, South Carolina"

The Carolinas mills developed from 1880, and would employ children in preference to adults. At Newton Mill, North Carolina, in 1909, twenty of the 150 workers observed, appeared to be twelve years old or less. As well as the usual report of hands and fingers getting severed by the machinery, and insufferable heat- the dust inhaled caused a fatal condition known as brown lung. Laws were rarely enforced, and the presence of small children in the factory was explained away to the inspectors saying they were visiting the mill to bring meals to their parents (meal totters), or helping but not on the payroll (helpers). Wages were good for the workers who could earn $2 a day in the mill against $0.75 on a farm. In the segregated south, 'Blacks' were not allowed to work inside a mill; had they been the need for child labour would have been eliminated. Child labour stopped here not only because of new laws but the change in the type of machinery caused by the Great Depression, which required greater height and skill.

Women

In 1926 when it was at its peak, the Lancashire cotton industry worked 57.3 million spindles and 767,500 looms. It imported 3.3 million bales and exported 80% of its production. 61% of the 575,000 cotton operatives in Lancashire were women, of which 61% were unionised in 167 different unions

Unions

The 18th-century woollen industry of small producers in southern England was far different from the Yorkshire-based worsted industry where the clothier imported and owned the raw material and sold the cloth. He put out work to small weavers, in effect, employing them. Worsted was more capital intensive. The small weavers banded together to form self-help guilds. When Lancashire adopted cotton, the same process occurred. But in Lancashire cotton mills, spinning became a male occupation, and the tradition of unions passed into the factory. As spinners were 'assisted' by several 'piecers' there was a pool of trained labour to replace any spinner the owner cared to dismiss. The well paid mule spinners were the 'barefoot aristocrats' of labour and became organised in the 19th century. They paid union dues, and were well placed to finance themselves should a strike be needed. The Yorkshire worsted industry, adopted the ring frame which required less skill. Worsted spinning was an occupation for young girls. Unionism did not develop in Yorkshire until 1914. In, 1913 figures show 50% of cotton operatives were unionised while only 10% of wool and worsted workers.

In Lancashire there were:

Occupation	Union members
Weavers	182,000
Cardroom Operatives	55,000
Spinners	23,000
Piecers	25,000

The spinners union, the Amalgamated Association of Operative Cotton Spinners had a federal structure with strong central leadership where control was in the hands of a small group of paid officials. Their dues were high, so the fighting fund was large and the officials were skilled in defending the complex wage structures.

HEALTH OF THE WORKERS

A cotton mill was not a healthy place to work. The air in the mill had to be hot and humid to prevent the thread from breaking: 65 F to 80 F and 85% humidity was normal. The air in the mill was thick with cotton dust, which could lead to byssinosis – a lung disease.

Protective masks were introduced after the war, but few workers wore them as they made them uncomfortable in the stifling conditions. The same applied to ear protectors. The air led to skin infections, eye infections, bronchitus and tuberculosis. The noise levels in a weaving shop, where the shuttles in 500 plus looms were being thumped 200 times a minute led to levels of deafness in all who worked there. The lubrication was carcinogenic and led to cancers of the mouth and cancer of the scrotum; known as mule-spinners cancer.

A mill worker could expect to work a thirteen-hour day, six days a week with two weeks off for the wakes week holidays in summer. Unsurprisingly, a series of Factory Acts were passed to attempt to ameliorate these conditions.

In the early days when the cotton towns were expanding rapidly, living conditions for the workers were poor. Badly planned housing was seriously overcrowded. Open sewers and shared privies led to diseases such as cholera; Manchester was hit by an epidemic in 1831 that claimed hundreds of lives.

PAUL-WYATT COTTON MILLS

The Paul-Wyatt cotton mills were the world's first mechanised cotton spinning factories. Operating from 1741 until 1764 they were built to house the roller spinning machinery invented by Lewis Paul and John Wyatt. They were not very profitable but they span cotton successfully for several decades.

DEVELOPMENT

In 1738 Lewis Paul and John Wyatt obtained a patent for a cotton spinning machine that for the first time used the principle of two sets of rollers travelling at different speeds to enable fully mechanical spinning. The patent outlined the two key developments that were later to underlie bothRichard Arkwright's water frame and James Hargreaves' spinning jenny, and made it possible for a single power source to drive more than one spinning machine. Wyatt envisaged "a kind of mill, with wheels turned either by horses, water or wind."

Using this technology, and with financial support provided by associates of the author Samuel Johnson, in the summer of 1741 Paul and Wyatt set up the Upper Priory Cotton Mill in Birmingham, the first mill to spin cotton "without the aid of human fingers". The mill "containing fifty rollers... turned by two donkeys walking round an axis" was not a commercial success, with Wyatt unable to enforce the levels of organisation and discipline that an operation on this scale demanded; Andrew Ure was to comment that Wyatt was "favourably placed, in a mechanical point of view, for maturing his admirable scheme" but "a gentle and passive spirit, little qualified to cope with the hardships of a new manufacturing enterprise". Two years after its opening the mill was described as being in a "pitiful state" and in 1743 Wyatt was incarcerated in the Fleet Prison for debt. Matthew Boulton was later to observe that the Paul-Wyatt mill "would have got money had it been in good hands", but nothing is known of it after Wyatt's release in October 1743.

Four other mills were set up using the Paul-Wyatt machinery in the following years. Edward Cave – publisher of The Gentleman's Magazine and one of the friends of Dr Johnson who had funded the development of Paul and Wyatt's invention – experimented inLondon with running the machinery by hand and by 1742 had set up 250 spindles at Marvell's Mill, a watermill in Northampton. This was the first cotton mill to be driven by water power. Cave experienced similar problems organising his workforce to Paul and Wyatt ("I have not got half my people come to work today" he wrote to Wyatt, "and I have no great fascination in the prospect I have to put myself in the power of such people") and the mill generated little profit, but was to continue operation until about 1764. Samuel Touchet set up a second Birmingham mill in conjunction with Paul and Wyatt in 1744; little is known of its fate, but it was sufficiently encouraging for Touchet to lease the Northampton mill from Cave for a period up to 1755.

The final mill was Pinsley Mill established by Daniel Bourn in Leominster. This may have been in operation by 1744, but is first mentioned in 1748, when both Bourn and Paul patented machinery for carding cotton - a premilinary process that must be undertaken before spinning. Bourn's mill burned down in 1754 but must have had a considerable reputation, as the Manchester Mercury's report of the fire described it as being "erected there with great expense and skill" and "viewed with great pleasure and admiration by travellers and all who had seen them"

3

Development of Conventional Textiles

The India of today succeeded to a rich heritage in Architecture, Music and other Fine Arts and Crafts, which for scores of centuries had been gradually growing despite the rise and fall in the destinies of its people. Aliens who came, whether as traders or as invaders, ended with the solitary exception of the English.

By making India their home and by getting absorbed in its original inhabitants. The mixing of races did begin new currents and cross currents in the flow of our culture, but it made our culture richer and fuller.

The culture, India invariably shared with its neighbours and others, thereby earning for itself an honoured place-second to none-amongst the nations of the world. Its fame spread to lands far and near, even in those remote ages of which no written historical record exists.

Whilst the missionary zeal of our religious emissaries carried the spiritual aspects of our culture to the distant corners of the globe, the spread of our culture on its temporal side was completely due to our traders and exporters who were our cultural ambassadors abroad.

Through the export of its reputed textiles the exquisite beauty and fineness of which was almost legendary, ancient and medieval India attained an unrivalled position in the international textile trade, which she maintained till not-so-very-long ago.

According to George Birdwood, India "was perhaps the first of all countries that perfected weaving and the art of its gold brocades and filmy smuslins". Can it be wondered then that the textile products art the Indian carftsmen were eagerly 'and even anxiously looked *for* and awaited in the markets of the world? How could competitors ever hope to catch up with a country that was thousands of years ahead of them in this art? They could only ban the utilisation of the Indian fabrics within their own country.

The Roman Senate was probably the first to pass a law (abxout 1600 A.D.) prohibiting the wearing of the Indian silk garments by men. A little more than a century later England prohibited by law the use of the Indian calico.

Fig. Weaving by Hand

This was the starting of the two-centuries-long unremitting war against India's supremacy in textiles. As is only natural, a crop of legends has collected around the fineness and the delicacy of. the fabrics which the Indian weavers produced with the aid of little more than bare human hands. One such, which is very believed to be authentic, might be related here.

Hand spinning and weaving though carried on a lower scale and developed gradually as a cottage industry had yet reached a high stage of perfection, even as early as 327 B.C. We catch glimpses of it in the descriptions, sent home by Alexander's soldiers, of the costumes worn by the people of India. Foreign visitors to this country, on their departure, always carried away with them, as priceless treasures, pieces of silks, brocades and muslins. Cargoes, laden with our wondrous fabrics, were invariably welcomed by other countries.

This marvellous technique of the craft which. has raised it to the status of a 'Fine Art' has evolved through hundreds of generations of craftsmen who formed an exclusive class or guild and handed down from father to son, the knowledge gained and the progress made so that each successive generation of craftsmen added something to it and became more and more efficient, producing belter and yet better results.

Another feature of traditional Indian textiles common to all artistic creations of human hands is that, each piece bears the stamp of the individuality of the craftsmen who produced it. Thus, it has a human interest of its own. As the gifted writer Shm. Kamla Dongerkery has said, the traditional textile of India "reveal the background of great culture, they give artistic shape and form to the ideas and ideals which inspire the lives of the people" and thus provide one of the *most* reliable hallmarks of the cultural development of the people.

Fig. Spinning

The craft was carried on in the villages of some of the important provinces of India, particularly around the coastal towns and cities. It was not restricted, as a profession, only to men alone, but women too, in their own leisure hours, helped their men in spinning and weaving and thus, added to the family income.

Indeed, in certain provinces like Assam, tradition needed women to achieve some degree of skill in the arts of weaving and spinning and even today unmarried girts in Assam are notn considered very eligible for matrimony, unless they have acquired some grounding in this useful art.

But the craft flourished mostly as a result of the patronage extended to it by the royalty and aristocracy, which played a significant role in its development. To suit the tastes of the royal and noble families, the standard of production had necessarily to be very high and ever progressive. Various kinds and varieties of textiles were produced in different parts of India and thus we hear of Benars and Sural Brocades, Dacca Muslins, Cashmere Shawls, Gujarat Patolas an Bandhanis, each reputed for its design, quality and colour. It would probably interest our readers to learn that in anum. ber of ancient families are cherished, even today, as valuable handlooms, some beautiful specimens of our conventional textiles. A good few of these have been presented to some national museums which proclaim to the nation and to the world of Fine Arts "the glory that was Ind".

Saris of Dacca

Upto the beginning of the 19th century, the Dacca muslin saris, one of the most artistic and beautiful specimens of hand-loom textiles, were counted amongst their valuable and cherished possessions by the women of Bengal. Even today, these saris are beautiful in their designs, but alas the art of making fine muslins is no more, and the saris too are not half as fine as they were in the past.

The saris are usually grey, white or black with blue or black designs. Occasionally, the patterns are woven in with bright coloured cotton, or silver or gold threads. The Dacca muslins with the woven in pattern are known as 'lamdani' and the typical designs of flowers or figures utilised by the Dacca weavers are called 'Jamdani' pattern. The saris have very bold and large lamdani pattern on the 'Anchal' 'Palloos' (end portion) and the borders. The rest of the sari is generally covered with several small *bootties.*

The common motif is the round design bootties, which suggest *Chameli (Jasmine)* flowers and around these are woven the leaves that recalrthose of the sweet smelling *champak.* When the sprays of flowers are spread all over the sari, it is called a lines, the sari is known as *'Terchha'.* But when the floral desigtt forms a net-work which covers the entire field, then the pattern is known as *'Jatar'.* Sometimes in *Jamdani* design, the flowers are clustered together-like the settings of jewels, and then the pattern is provided the poetic name of *'Panna hazare'* (thousand emeralds).

If a running floral pattern covers the whole field, the expression Phulwar is utilised; but if the flowers are large and life-size the Jamdani is caused Toredar. The borders and Palloo or Anchal (end portion) of saris are mainly decorated with distinctive figure designs. The figures chosen represent birds, animals, and human beings.

Peacocks or *'mayura'* and herons or *'hansa'* seem to be popular as bird-figures in the designs of Dacca saris. Also some of the motifs indicate the influence of mythological legends, as well as of the local traditions. The designs are commonly accepted as of Persian origin but several of the designs depict incidents from the Hindu mythology also.

The most striking feature of the *Jamdani* pattern is the skill of the weaver in being able to depict the conception of actual motion in the figures he weaves in. Several a design presents birds with flapping wings as if they are about to flyaway. The outlines of figures are always bold, straight and invariably geo. metric in clean cut lines.

Also the intervening space is so well balanced with lines and flowers that a most delightful effect is produced by the combination. It may well be imagined that the weaving of such masterpieces called for the highest skill and crasmanship with almost unlimited patience as each design must have developed months of work. An idea of the skill of the weavers may be formed from the following description of the process of weaving.

Clothes of Muslins

Dacca (now the capital of Eastern Pakistan) was, for centuries, synonymous with the finest muslins the world has ever produced by hand or machine. In the words of Dr. Forbes Watson "with all our machinery and wondrous appliances we have been hitherto unable to produce a fabric which for its fineness and use can equal the 'Woven air of Dacca'." Indeed, the Dacca weavers' Imagic hands produced such exquisitely fine and delicate fabrics that three poetic names *'Ab-i'rawan'* (Flowing water), *'Baft-Hawa'* (Woven-air), *'Shabnam'* (Evening dew) werel justifiably given to them.

Exhibits in some of our museums 'Prove even today that a yard's width of the muslin could easily pass through a lady's ring. Then there are legends which have gathered around these fabrics. One of them relates that a five-yard piece of the muslin could be packed in a match box. It is, however, an authenticated fact that a 15 yard piece of 36 inches width *'Mulmul Khas'* (Royal muslin) in the reign of the Emperor Jehangir, weighed only 900 grains. Of course *'Mulmul Khas'* was the Dacca muslin.

The value of Dacca muslins is estimated by the number of warp-threads in a provided length of the material as compared with its weight. The greater the length and the number of the threads, with relatively less weight, the higher would be the price. For example, a yard's width of *'Mulmal Khas'*, the finest muslin, was known to have 1000 to 1800 threads in the warp, and the weaver took five months to weave 10 yards of the fabric. The weaving of these fabrics could only be done during the rainy season, because for the weaving of such an extremely fine fabric, a humid atmosphere was essential.

Saris of Chanderi

The muslins woven in *Chanderi,* a place for themselves due to near Gwalior, have earned a name their fine quality. *Chanderi* saris are mainly cotton with borders and palloos woven in silk or gold threads. Sometimes mixed threads of silk and cotton are used for weaving the fabrics known as *'Garbhreshmi'.*

The palloos of these are very artistically ornamented with gold threads while the ground of the sari is checked, with *bootties* in the centre of each check-square. The borders are woven with double threads which produce an effect of two colours, one on each side. The saris are woven in nine-yard lengths and are very much valued by the Maharashtrian ladies.

Buttedar of Baluchar

Baluchar, a small town near Murshidabad, has become a noted and a highly valued name in the handloom textile history of India. The artisans of the locality produced very artistic figured silk saris known as Baluchar Buffedar. In Baluchar Butfedar, as in Dacca saris, the Pallo os were the most elaborately ornamented portions.The field of the remaining part of the sari was decorated with small bootties of some floral design or figure design of birds.

The special feature of Baluchar Buttedar is that the design used for the ornamentation display a strong influence of Mughal Art, which is, famous fore its portraits. The weaver of Baluchar chose motifs of human figures and the popular 'Toranj' (also called 'Kalka' or 'Guldasta') which is the most popular motif in weaving, embroidery and printing throughout India under its present appellation 'the mango design'.

An illustration of Baluchar buttedar displays the richly ornamented palloo and a portion of the field of the sari. In the design of the palloo, the famous ever popular 'Toranj' are seeri as though these are set in a frame. The border of the frame is again elaborately decorated with pictorial representation of a lady smelling a flower and stated in a sort of a nich.

The interspaces are filled with clearly arranged rows of *'Toranj'* lined with an outer border of flowering plant. The border design which is a simple and straight combination of a little *toranj* and flowering plant is continued for the border of the whole sari. The subjects for portraits were either a lady or a nobleman dressed in Persian dress and holding a flower or riding a horse or smoking. Though the subjects were always of Islamic origin, yet the *Baluchar Buttedar* were very fancies amongst the Hindu ladies.

This gesture indicates a complete absence of intolerance existing between the two communities. Later on, probably due to the desire for the patronage of the English bosses of the East India Company the woven-in-pictorial subjects came to include figures dressed in European clothes, and holding instead of the traditional flower a wine glass. The significant art of weaving figured fabrics in Baluchar is lost for ever and a few extinct scattered specimen in some museums are the sad mementoes of the perfection it had achieved.

Barta Kam Khwab, and Abi-rawan (Brocade)

Kam Khwab is the name given to real gold brocades. 'Brocades' is the expression used by Westerners who at first caued it 'Kin Kab' or 'Cin Cob'. According to their conception, brocades are thick textiles with woven in pattern very prominently thrown up on the surface of the face of the cloth. Consequently, not only Kalil Kawabs, but other similar textiles also like Buftas, Amrus, and Himrus are known to them by the wide English name 'brocades'.

The poetic name Kam Khwab expresses the dream-like beauty and richness of the fabric. It literally implies only a little less (Kam) than a dream (Khwab) or a dream (Khwab) reduced (Kam) to reality. The real Kam Khwabs are woven with pure gold threads and the silk yarn is added to provide a bJdy and as a means for colour illuminations. Silver or gold-plated silver threads are also utilised for keeping down the price but gold is usual. These are heavy fabrics and are generally used for making palloos, blouses and for men's halfsleeve Indian jackets, long-coats (Angarkhas or Achkans), ceremonial robes (chogas) and also later on came to be utilised for curtains and for upholstering the furniture of Public Rooms and Durbar Halls of Princes.

Thans of Pat or Bafta

These are the brocades in which the major portion of the fabric is in closely woven silk with patterns in gold or silver at regular intervals. This fabric is usually woven in a narrow width of 20" to 30". This is not as heavy or thick as Kam Khwabs but is heavier than the other textiles. Baftas are also utilised for blouses and Indian skirts (Lehngas or Damans) and for men's Angarkhas or Ackhans too. Ab-i-rawans. These are silk gauge materials with gold or silver patterns only on certain portions.

Silk gauze ceries with woven-in gold borders and palloos, are also called Ab-i-rawans. The literal meaning of this poetic name is 'flowing water'. Although brocades are a speciality of Banaras for which it famous yet these are also manufactured at Surat and at Ahmedabad displays an Angarkha made in Ahmedabad brocade. The design in the illustration depict the traditional floral motif. In old brocades motifs representing animals and human figures enclosed within floral borders were in use.

PITAMBARI AND PAITHANI

Himrus and Amrus Sari

Himrus are the famous silk brocades of Hyderabad (Deccan). The State's second largest town-Aurangabad is the chief centre of the art of Himru-weaving. Himru perhaps a derivative of the Sanskrit Him (snow) is a fabric utilised in winter. The ground is cotton, and silk is used for the brocade on the surface. The yarn used for weaving Himrus is spun so as to produce, when woven, the effect of a warm soft material like wool.

The peculiarity of the *Himru* is that the silk thread which is used to form a pattern on the surface of the cloth, is carried to the reverse side of the cloth and is collected there in clumsy long loops. This forms a rather loose but soft warm layer. Further, the accumulation of the loose threads on the reverse of the cloth, necessitates a lining to all garments made of *Himru* cloth. Thus *Himru* garments make very warm clothing suitable for the cold season.

When silk thread is used exculsively for weaving *Himru,* the fabric is called *'Amru'. Amrus* are generally made in Ahmedabad, Surat and Banaras. *Himrus* are used for men's *Achkans, Chogas,* and for female wear also, *e.g.,* for blouses and *Lehngas.* For generations, the Nawabs of Surat used a special quality of *Himru* fabrics for their dresses which was called the *'Nawab's Himru'.* These fabrics are also used for upholstery and curtains.

Pitambari and Paithani Sari

Paithanis are the beautiful and rich saris made at Pattan or Paithan in the state of Hyderabad (Deccan). These are exquisitely fine fabrics with gauze like texture ornamented with gold patterns woven in the texture of the cloth. The borders and the palloos which are woven separately as gold brocades are sewn on to the sari.

The colour of the sari is generally dark orange, red or yellow, with gold lines arranged *in* a check or in stripes. The inter-spaces are usually filled In with a figure design depicting a goose with an olive branch in its beak. The borders and the *palloos* have very striking designs in bright and showy colours such as moss-green, canary yellow, and bright pink. The common motif of the designs is the peacock supporting a big vase with sprays of brilliantly coloured flowers so arranged as to form a Persian cone pattern.

The illustration of a *Paithani palloo* does not have the favourite peacock in the design, but it depicts the harmonious arrangement of the sprays and the surrounding floral design. The vase with sprays is placed between two pillars joined with the *toran* (arch). The design is worked in silks of blue, red, and white colours on a field of pure translucent gold. The whole effect is gorgeous and is eminently artistic in its perfect harmony.

In olden days *Paithanis* were usually woven to order for the Royal family and the weaver took months to complete a single piece. The value of a genuine real *Paithani* ranged between Rs. 2000 to Rs. 3000. In modern times, however, such highly valued and gorgeous *paithanis* are not woven. *Pitambars* are bright coloured silks 5 yards in length with gold borders sewn on them. These are worn by men particularly when performing any of the religious rituals.

Patola Sari

An artistica1ly ort Jamented fabric. It is a specimen of wonderful combination of the craft of tie-dying *(Bandhana)* and weaving. *Potola* is mostly in use as a wedding sari in Kathiawar and Gujarat.) In Java and in Indonesia too the Potola fabric is used for wedding dresses. The fabric is so exquisitely and very valued that it is handed down from generation to generation in the family.

Women of Gujarat and Kathiawar treasure the possession with pardonable pride. Patola, unlike the other ornamented fabrics, is invariably woven in just the plain weave. The elaborate and intricate patterns which mark the Patola saris are produced by the wonderful art of Bandhana or tie-dyeing. The silkyarn with which Patolas are woven, is first dyed by the Bandhana process before it is put on the loom. The yarns, both warp and weft, are dyed in the lightest of colours. Then they are stretched on the ground, and the dyer proceeds to mark some portions to indicate the lines of the required design. His wife who helps him in his work, then ties up the marked portions with cotton thread so tightly that the next dye cannot penetrate through to the tied portions. The yarn is then immersed in dye-baths of the desired colours and shades. The operation of tie-and-dye is repeated several times until all the colours and shades required for the planned design have been used to the yarn.

The dyer starts with a light colour, passes next to a bright one and applies the dark colour at the very last. Then the weaver starts on his job. The warp threads are arranged in pre-arranged sequence, and the weft is then interlaced one by one very carefully to form the planned design. The process of producing a *Patola* is, therefore, a very laborious one and is very complicated too.

Meticulous care and a good deal of creative imagination is required for marking the correct portions on yarns for dyeing in different colours of the pattern. A very retentive memory is another necessary requisite for registering and recalling accurately the sequence of the coloured threads in the pattern. Thus only a few old designs are utilised for *Patola* patterns. The following eight are used by the weavers of Pattan as described by Mr. G.V. Patel:

- Okhar bhat or water creat design. The real name of this design on investigation at pattern, appears to be akhrot bhat, i.e., the walnut design.
- Wagh-Kunjar bhat or tiger elephant design. The animals alternate with each other in the design.
- Chabri bhat or basket design. Here each enclosure containing an elephant is made up of four quadrants which look as if forming a basket when two of them are taken together.
- Nari-Junjar bhat or dancing girl and an elephant design. It has necessarily a parrot comprised in it.
- Chowkhadi bhat or a diaper with a double outline design. Each disper comprises three flowers borne on a stem."
- Pan bhat or leaf design. It is said to be the leaf of the sacred pipal tree. (Ficus Religiosa).
- Rattan chawak bhat or the cross of diamonds design. It has interspersed diamonds also.
- PhultJadi (hat or floral design) is ganerally enclosed in diapers outlined by a single line. Each diaper comprises three flowers.

There is one more design which is utilised for dhoties (the loin cloth worn by men). This design comprises of the devnagri alphabet and the forms of the letters follow those of the mantras (hymns) in religious book. Pattan, a place in Kathiawar, is reputed to be the birth-place of Patola.

The weavers of Pattan later migrated to Bombay, Ahmedabad and Surat and the making of Patolas started at these places also. Orissa weavers also have adopted the Patola technique for weaving their special fabrics suchas curtains, bed spreads, odhnis (scarfs worn over the head and draped round the shoulders and waist by women) and saris. The famous Sambalpore saris are woven like Patolas.

Bandhanis

Bandhans or Choonaris are the colourful saris and odhnis dyed by tie-and-dye process. These are fancied amongst the women of Gujarat, Kathiawar. Rajputana and Sindh. Premlata Jayakar in her article on 'Tie Dyed Fabrics of India in "Marg" refers to Bandhanis.

It is an auspicious garment. A symbol of youth and romance, love play and the 'Sohag' (wifehood) of Hindu women. It is a garment of laughter'. Indian women are known for their love for bright colours. Also the tradition and the customs of wearing special colours on different festivals, makes it necessary for them to become familiar with the art of dyeing at home. Thus besides the expert professional dyers almost every Indian girl learns by practice a good deal of the art of dyeing and Bandhani work.

Bandhanis differ from Patola as regards the stage at which they are dyed. Like Patolas they are dyed by the tie-and-dye process, which, however, is done after the fabric is woven. The fabric is folded over many times until reduced to a small thick square or a rectangular piece. The piece is then damped and pressed on a block on which a design has been carved. The impressed portions are picked up by the finger nails (the nails are allowed to grow particularly for the purpose and are used as a sort of pincers) and are then tied up with cotton thread in a thickness sufficient to resist the dye. It requires training and great skill to pick up all the layers at once and make it crinkle in a particular given manner.

The Bhalldhanari or the woman who does the tieing up work works swiftly and ties up all the impressed portions without cutting the thread but carries it over from one point to the next. The dyeing process is carried out in the same order as in Patolas, beginning with the light colours and finishing with the dark ones. But each time, before a new shade or colour is applied the tieing up process has got to be repeated. generally, the designs used are copies of udarl worn by a Gujarati girl few traditional ones and y the practice1tieing up the same design over and over again the Bandhanaris become expert to such an extent that they are able to dispense with the process of impressing the fabric with the design.

The motif of the traditional designs used for Bhandhams reo presents animals, birds, flowers and dancing dolls. When "elaborate designs are used the Bandhanis are called 'Ghaghra chola'. In some of the costly 'Gharchola' gold threads are woven in to form checks or squares, and then the designs are' formed in each of the squares by the tie-and-dye process.

The 'Choonaries' are very light fabrics, and the designs for these consist of dots or pin heads irregularly spread all over the sphere of the cloth. Seldom the dots are grouped together to form a design, and the design is known as 'Ek bundi' (one dot), 'Char bundi' (four dots) and 'Sat bundi' (seven dots).

It might interest our readers to know that in some parts of Rajputana, e.g., Alwar, professional dyers existed till a couple of decades ago, who could dye even the finest muslim in two different colours, one on each side of the fabric at the modest charge of only annas four a yard. This art too is now extinct but specimens can be found in some museums.

KALASDAR FABRICS

Also known as Palampores in the textile trade. They are available in rectangular pieces and are popular with Hindus and Muslims alike. The former use them as cano-pies for the images of their gods and the latter as praying carpets. Those designed for Hindus portray scenes from the Hindu mythology, whilst those intended for Muslims are of Islamic origin.

The latter depicts the conventional Muhab with panels forming a frame enclosing the 'Persian Tree of Life' complete with birds and the branches and animals resting under its shade. The craftsmanship and the skill of the dyer is amply borne out by the excellent portrayal of the minutest details with amazing accuracy.

The French traveller Bernier who visited India in 1663, during the reign of Emperor Shahjehan, thus describes the fabric which formed the drapings of the Imperial courtyard.

EMBROIDERED FABRICS

India is rich in embroidered fabrics. the Indian art of embroidery is an ancient one. Its origin can be traced right back to the Vedic age and it appears to have been well-developed by the time of the great epics. Later on, the influence of incoming races and tribes particularly those which migrated from the hills contributed a great deal toward the development of the art along lines which have given it, its present form.

"The stitches employed and art conceptions displayed" by these early artists indicate the extent of their knowledge of the art. It may be noted that throughout the mountains and valleys of India, the art is very popular and is assiduously pursued by men and women. The very colourful embroidery produced is not always intended only for the market, but for home use as well.

Kabnendar or Kalarndar

This is the name given to the hand painted cotton fabrics. They are so called because the artist works out the designs on the material with a fine steel brush not unlike a pen (Kalam). The process is very much the same as used for Batik work. The basic principle, namely, resist-dyeing, being common to both.

The material is first dyed in pale tpink and then stretched out tight. The artist then traces the outline of the design with his Kalam or fine steel brush dipped in melted wax. The fabric is then dyed deep red and finally washed in hot water to melt away the wax. This produces the design in deep red on the background of pale pink.

And lined within with those chittes, or cloth painted by a pencil of Masulipatam, purposely wrought and contrived with Palampore such vivid colours and flowers, so naturally drawn, of a hundred several fashions and shapes, that one would have said it was a hanging parterre. A few of the better-known embroidered fabrics may now be briefly described.

Shishedar Phulkaries

These are different from the other phulkaries. The special feature is that the embroidery is embellished with tiny mica or looking-glass (Shisha) discs fixed to the cloth by button-hole stitches, The material embroidered in this way is very often silk, or even satin. Sindh specialises in this kind of phulkaries and is reputed to be its home.

The Punjab Phulkari-Phulkari

Really means flower (Phuf) work (kari). These were the conventional ceremonial shawls worn by the Hindu bride at her wedding when going round the. sacred fire with the bridegroom. It is a popular saying in Punjab that when a girl is born in a family the mother, or may be the grand-mother, starts embroidering a Phulkari to be presented to the girl on her wedding day. The women of the Jat community are specialised in Phulkari work. In almost all the districts of the Punjab, wherever this community settled down the Phulkari work originated and has flourished. The peasant women of Rohtak, Hissar, Gurgaon and Karnal are known for embroidering the best Phulkaries. Rohtak is considered to be the home of Phulkari work.

The stitch for Phulkari embroidery is the simple darn stitch. The materiaL on which the embroidery is worked is rough Khaddar, but the thread used for the embroidery is silk and the colours used are white, red, yellow or;green". The colour of the Khaddar material is invariably red, maroon or brown. The Phulkari motifs generally of floral patterns, but geometric patterns are also used.

For instance in the Bagh or Shalimar design, the entire material is covered with a geometric pattern. In orthodox Phulkari, however, the same floral pattern is embroidered only at intervals on the cloth, intervening portions of which are left plain. Sometimes, only the borders"are embroidered to a width of 3 or 4 inches and the centre of the material is left plain. This pattern is known as chobes. The choice of the design is wide, but very often they appear to be almost identical becauce of the close and compact stitches used.

It is remarkable that the illiterate village women choose designs and the colour schemes which have a charm and beauty of their own and which are worked out unerringly by memory.

Silk, or Even Satin

Sindh specialises in this kind of *phulkaries* and is reputed to be its home. The designs used indicate the influence of the Punjab as well as Cutch in the kind of stitches used and the colours selected The phulkari work of Upper Sindh appears to follow Punjabi Phulkari while that of the Lower Sindh bears a resemblance to the Cutch work. Besides the, button-hole stitch, the darning stitch of Cutch are also used in the Shishedar hulkades.

Cutch Phulkari

This is a name given to the embroidered silk or satin material used for skirts in Cutch. Illustration of the Cutch *Phulkari.* The ower portion of the material is embroidered to form a border, and the rest of the material is filled with floral or figure designs. The popular figures chosen by the Cutch artist for embroidery are elephants, peacocks and parrots. Most of the embroidery is done in satin stitch, but occasionally herring bone stitch is also used for finishing the edge of the border.

Roomals of Chamba

The Roomals of Chamba, a state in the Himalayan range, are important pieces of embroidery. Princesses as well as shepherdesses are equally agept in the art which they have adapted as a pleasant or profitable pastime for their gap hours. The Roomal is a square piece of botton material of the size of a teapoy cover. The material usually selected is very fine and of delicate texture. The embroidery is worked in double satin stitch which produces the design on both sides of the material and so a Chamba Roomal does not have a right or a wrong side.

The motifs for designs are figures, flowers, leaves, forming a sort of frame to set off the central representation of a scene from a mythological story ora legend. It is said of that embroidery that it "imitates the Pahari School of painting". The unique feature of the embroidery is that it gives a vivid impression of the embroidered figure being in action or in motion, thereby enhancing the artistic value of the Roomals.

Kanthas Fabrics of Bengal

Kanthas are not original fabrics, but are made with lengths of old and practically used up cloth. Many pieces of about the same 1ength are placed one on top of another and the edges of all the pieces are sewn together so as to form a padded or quilted rectangular piece. Then the artist proceeds to depict beautiful patterns or scenes from stories from the epics or famous legends, by means of embroidery done in simple running stitch. The work is very fine and neat and properly executed completely from memory without the help of any tracing or drawing or any written notes.

The *Kanthas* are therefore classed as works of art and one has only to look at them to agree. The women of Bengal generally devote their leisure hours in working *kanthas,* creating beautiful and artistic fabrics out of worn out clothes. Embroidered Fabrics of Kashmir. Embroidered fabrics of Kashmir are world famous. The Pashminn shawls, the silk saris, the Namdas and the various other silk and woollen artists are praised as works of art. The Kashmir embroidery is known as *Kaseeda,* and the stitches used are the satin stitch, the stem stitch and the loop stitch. The darning stitch is also used.

The nerring bone stitch is used for the edges of the finished pieces. The craftsmen, in Kashmir, are generally men and are in variably helped by boys, often of a tender age, who do the actual embroidery. It is interesting to observe the young boys at work. The master craftsman calls.

The design before him, the kind and the number of stitches to be put in. As the instructions are called out the boys work swiftly and deftly with needles and the stitches are completed almost as soon as the masterhastinished calling them out. The scene similar to that of a small, class in a school with the boys taking down the dictation of the teacher.

Nature's bounteous gifts appear to have been literally showered on Kashmir and it is considered by many to be the nearest approach to "a paradis on earth". Kasheeda artists, therefore, under the inspiration of the beauty of the natural surroundings which they have succeeded in reproducing in their embroidery with such amazing skill, that in the words of Shm.

Kamla Dongerkerry, "competes with the wealth of Nature's charms". The Kashmir artist has an inexhaustible treasure of, motifs in gorgeous colours, ready at hand from which he can draw his designs. In Kashmir embroidery, therefore, we find bunches of fruits, foliage and birds of brilliant hues and-innumerable shades of colour lend themselves as charming compounds of different designs.

And the *Turanj* (Of the mango) is not ignored either. Indeed, it finds a place in almost all the *Kasheedas* of Kashmir. The outstanding features of Kashmir embroidery is its elegance and the harmonious grouping of the brilliant colours in the design which produces a restful and soothing effect.

The 'workmanship is so neat and the stitches so fine that one wonders how bare human hands and that too of young boys could produce such work. It must, however, be remembered that the training of generations (the art is handed down from father to son in each of the families) has equipped the artisans with almost an intuitive aptitude for the work which has become second nature to them. On the other and, the art lives. only so long as the line of the craftsmen continue and is likely to become extinct with the line. This, alas, is what began to happen some time ago due mainly to the rage for replacing human hands by lifeless machinery. Paradoxical as it might see science is fast stifling and killing art by restricting the scope of arts creation only to such articles as lend themselves to huge production by mechanised processes. The modern products of Kashmir are no longer as fine and delicate asin the past, and tend to lose their natural charm. The embroidery is best done in silk but the similar patterns are also repeated in wool. Besides the Pashmina shawls arid the silk saris, the industry produces many articles for use in the home. Namdas are the embroidered rugs.

The rug is a felted thick fabric manufactured by the process of pressing wool and cotton together. They are then embroidered in thick wool in bright colour designs similar to those on Pashmina. Kashmir Silk embroidered fabrics besides bring colourful and rich in design and comparatively inexpensive, are within the means of middle dass people. Thus wider patronage supports the industry and the art has nourished not by the patronage of the great and the rich alone.

Chikankari Embroidered Muslins

The white embroidery on white cotton particularly on muslins is called Chikan work Chikankari is an industry nurtured and developed in the region watered by Ganga and her sister Yamuna (Jamna).

Lucknow in especial is the cradle and unrivalled centre of the art. Chikan work lack in the attractiveness associated with colour yet has a fascination of its own, unequailed by few and surpassed by none. Whilst the kaseeda of Kashmir may be rich in colour, reflecting in silk the ravishing beauty with which nature has gifted the valley the virgin white chikankari is perhaps a translation in shimple cotton of, the purity of the waters of the sister rivers born in the home of regular snows the Himalayas.

Much credit for the high position of the art of the Kashmir embroidery goes to its attractive colours. 'The fact that without that colourful aid, the *chikankari* has attained the same eminence, is eloquent of its great artistic value.

Daintiness and delicacy (which are the hall-marks of al1 Lucknow works of' art) added to a firtish and a richness of its own, are 'the outstanding features of *chiktmkari.* The work is sometimes so fine that to the naked eye it presents, the appearance of having been woven-in, along with the fabrics.

There are two chief styles of *chikan* embroidery:

- The flat—this group includes the *bukhi* and the *katao* styles.
- The knotted or the embossed, of which the *murri* and the *phanda* are the well-known varieties. To these may be added a third style, namely, the *jali or* netting which is akin to the drawn thread work, but is produced in completely different 'way.' The drawing out of

threads, is, regarded slovenly by the jali embroiderers.

'Instead, the fabric 'is pieced with holes of the requisite size to suit the pattern and these are then tied up to produce an appearance of net. *Jalis,* are of different kinds variously named *Chennai loli, Kolkata Jali,* etc. and are very elaborate and intricate. The kinds of designs are said to be thirty in number. Considering the fact that the colours are not used to vary the designs, and only the forms and motifs alone produce the different patterns, the number is creditable. The design represents familiar objects connected with daily life, often grains suchas rice and millet, in a variety of combinations.

'The *katao* produces an effect similar to *bukhia.* The save fabric is used to produce an opaque effect and for the outline' a simple prdinary stitch is utilised.

The *bukhia* is the most intricate as it is the most remarkable of *chikandesigns.* It is supposed to be the true *chikan.* Shri mati Kamla S. Dongerkerry considers its comparable to the shadow work of the present day" and thus describes the technique of its production. The stitches in this design cover the back of the cloth in the style of the herring-bone stitch, producing an opaque effect on the surface of the fine white fabric and at the same period an outline of motifs of flowers and leaves with minute stitches resembling the strokes of the back stitch.

The murri and the phanda are used mostly in the patterns representing grains. Chikankari work is mostly used for sari borders, blouse or kurta collars and tm recent times it was also utilised for the neck pieces the cuffs, and even the hems 'of men's angrakhas (long, coats). It has lately come into use as table linen, tea-cosy covers, and numerous other washable articles of domestic' use. 'Perhaps due to' the influence or the Western customers who invariably look for colour.

TEXTILE INDUSTRY OVERVIEW ON INDIAN CONTEXT

Textile Industry is one of the largest and oldest industries in India. It has a significant role in India as it fulfills the essential and basic need of people. Textile Industry in India stands at unique place and has maintained a sustainable growth over the years.

This is a self-reliant and independent industry and has great diversification and versatility. Textile Industry in India provides great contribution for the development of economy. It is the second largest textile industry in the world after China. It provides ample employment opportunities to people belonging to all classes. After agriculture this industry provides employment to maximum number of people in India employing 35 million people. Textile Industry represents the rich culture, tradition, heritage and economic well-being of country with diversified range and versatility. At the same time industry is competitive enough to fulfill different demand patterns of domestic and global markets. Indian Textile Industry plays vital role in country's economic development and contributes 14% to industrial production in the country. Textile Industry contributes around 4% of GDP, 9% of excise collections, 18% of employment in industrial sector, and has 16% share in country's export. Indian Textile Industry is valued at US $ 36 bn. The development of Indian Textile industry started in 1985. This was the year when, for the first time textile sector was considered as an important industry and a separate policy was formulated for sector's development. In the year 2000, National Textile Policy was announced. With further development Textile Industry came out of Quota Regime of Import Restrictions under the Multi Fibre Arrangement (MFA). This development came on 1st January 2005 under the World Trade Organization (WTO) Agreement on Textiles and Clothing. Because of the elimination of quota restrictions, most of the developing countries now can develop the potential market at both domestic and international level. These countries can develop the industry expertise and can have competitive advantage through implementing new technology, more skilled labour,improved distribution channel, cost effective operation and production with greater value addition in each step of value chain. Moreover it will help for Foreign Direct Investment in industry that will create great opportunity to strengthen the sector.

Some of the strengths of Indian Textile Industry are large and potential domestic and international market, large pool of skilled and cheap labour, well-established industry, promising export potential etc. There is a huge demand of Indian textile products in international market because of its versatility and it is expected that industry would grow with rapid rate in future and contribute towards economic development of the country handsomely.

Indian Textile Industry is one of the largest and oldest industry in India:

- Indian Textile Industry is highly fragmented industry; at the same time it is an independent and self-reliant industry that has shown sustainable growth over the years.
- Indian textile Industry is valued at US$ 36 Billion with exports of US$ 17 Billion in 2005-2006.
- Indian Textile Industry is second largest industry in terms of providing vast employment opportunities and employs around 35 million people in country after agriculture sector..
- The Indian Textile Industry plays vital role in economic development and contributes 14% to industrial production in the country.
- Textile Industry contributes around 4% of GDP, 9% of excise collections, 18% of employment in industrial sector, and has 16% share in country's export.
- Industry has direct and strong linkage with rural and agriculture sector, therefore it is estimated that, one of every six households in country is directly or indirectly dependent on this industry.

- Industry contributes around 25% share in the world trade of cotton yarn.
- India is evolved as a major contributor in world's cotton sector. Indian is the world's third-largest producer of cotton and second-largest producer of cotton yarns and textiles.

JUTE TEXTILE SECTOR

Jute Sector plays very important role in Indian Textile Industry. Jute is called Golden fibre and after cotton it is the cheapest fibre available. Jute, the golden fibre has gained immence popularity around the globe because of its biodegradable character.

Indian Jute Industry is the largest producer of raw jute and jute products in the world. India is the second largest exporter of jute goods in world. This sector is one of the major source of employment and supports the livelihood of 40 lakh farm families and 4 lakh workers.

The total Jute Mills in India are 78. Out of these 61 in West Bengal, 3 in Bihar, 3 in Uttar Pradesh, 7 in Andhra Pradesh, 1 in Assam, Tripura, Chattisgarh, and Orissa. The exports of jute goods are between Rs. 1100-1200 crores. The production of raw jute varies between 90-180 lakh (180 kg. each). The Domestic consumption is 13.5-14.5 lakh MT. In 2005-06 the production was 85 bales and estimated production is 105 lakh bales for jute season 2007. For grater growth and development of the Jute Industry Govt. of India launched National J Trade Policies

TARIFF POLICY

India and US have reached on an Agreement for reciprocal market access commitments for Textiles and Apparel with the negotiation of the WTO Agreement on Textile and Clothing. It provides elimination of Quota system of Textiles and Apparel from 1st January 2005.

- Under Indo-US Agreement of 1st January 1995, India agreed to reduce tariffs on Textile and apparel and remove all the restrictions on these products.
- From 1st April 2000, Govt. Of India reduced tariffs on: Manmade Fibres and Filament Yarns from 35% to 20%
  - Cotton Yarn from 25% to 20%
  - Spun, Blended, and Woolen Yarn from 40% to 20%
- India agreed to bind its tariffs on 265 textile and apparel products (Textured Yarns of Nylon and Polyester, Filament Fabrics, Sportswear, and Home Textiles.)
- Apparel products are free from Excise Duties and various Taxes.
- Grey Fabrics and certain Cotton Yarns are exempt from basic Excise Duty.

- Customs duty on Polyester Filament Yarns is reduced from 10% to 7.5%. Duty on other Filament yarns will be remain at 10%.
- Customs duty on Polyester Staple fibres is reduced from 10% to 7.5%. Duty on other Man Made Staple fibres will be remain at 10%.
- Customs duty on Raw Materials such as DMT, PTA and MEG reduced from 10% to 7.5%.
- For Small Scale Industries there is Full Exemption Limit being increased from Rs.1 crore to Rs.1.50 crores.
- Most of the products fall under HS code 61 and 62 carry an import duty of 56.83% which includes 30% basic duty, 16% additional duty and 4 per cent special additional duty.
- Excise duty on Nylon Chips has been reduced from 16% to 12%.
- Optional excise duty on Nylon Fish Net Fabrics is increased from 8% to 12%.
- Excise Duty Exemption on specified Textile Machinery Items is withdrawn and 8% Excise Duty is imposed.
- CST rate reduced from 4% to 3% with effect from April 1, 2007.
- Removal of surcharge on income tax on all firms and companies with a taxable income of Rs.1 crore or less.

IMPORT LICENSING:

India has liberalized its Import regime for Textiles and apparel, but some of the part is still limited for market access. Currently, there is no import restriction for yarns and fabrics items. Apparel and Made-up textiles goods require a Special Import License (SIL). Govt. revised Exim Policy on 31st March 1999 by eliminating Import Licensing Requirements for 894 consumer goods, agriculture products and textiles. On 28th December 1999 India and Us signed an Agreement for the elimination of import restrictions of 1,429 agriculture, textiles, consumer goods and apparel. India removed restrictions on 715 tariff items as of 1st April 2000.

CUSTOM PROCEDURES:

Marking, Labeling, and Packaging Requirements: Marking, Labeling, and Packaging Requirements for Textile products are technically complex and difficult to implement. According to textile regulation passed on 22nd july 1998 by GOI, Yarns, and Fabrics to have the statutory markings and these markings should not mislead the consumers.

For instance, Cloths must be remarked with the name and address of manufacturer, a description of cloth, sort number, length in meters and width in centimeters, and washing instructions. The Man made fibre cloth must indicate whether it is made by spun or filament yarn. The month and year of packaging, the exact composition of cloth. The Marking must appear on the face plait of each piece of cloth. The language for marking must be in Hindi and English with international numerals.

EXIM POLICIES:

- *Duty Entitlement Passbook Scheme*: DEPS is available for Indian Export Companies and Traders on a Pre-Export and Post-Export basis. Pre-Export credit requires the beneficiary firm has exported during the preceding 3-year period. The Post-Export credit is a transferable credit that exporters of finished goods can use to pay or offset custom duties on imports of any unrestricted goods.
- *Export Promotion Capital Goods Scheme*: This scheme is available to export companies and traders who provide the GOI with information about which type of goods and what value of Capital Goods they will import. And they also inform what will be the outcome of export they expect to produce from those imports. Depending upon the export commitment GOI provides them a license to import capital goods duty-free or preferential rates of duty.
- *Pre and Post Shipment Financing*: The Reserve Bank of India provides Indian Exporters Pre-Shipment Financing through commercial banks for purchasing raw materials and packaging materials by presenting Letter of Credit. RBI also0 provides Post-Shipment Financing through commercial banks at preferential rates by presenting export documents.
- *Export and Special Economic Zones*: Govt. of India has established Export Processing Zones (EPZs) and Special Economic Zones (SEZs). In EPZs units can import goods free of custom duty. There is 5-year tax holiday to any industrial unit in EPZs. Govt. has allowed 100% Fore3ign ownership of units under EPZs and SEZs. The Govt. considers SEZs as foreign territory for trade and tariff purpose. Units under SEZs may engage in Manufacturing, Trading and Services. Units are exempt from routine checking of exports by customs, and they can sell in the domestic market on payment of duty as applicable to imported goods.
- *Duty Drawback Scheme*: The basic objective of this scheme is to reduce the indirect taxes on exports. Exporters can get refund of the excise and import duty. Through this scheme they can be more competitive and have more potential market.

TEXTILE TRADE AND EXPORT

Indian Textile industry is one of the major industries that plays vital role for exports and foreign revenue.

- The share of Textile Industry to country's total exports is 16.63%.

- India's contribution in global textiles and apparel market is 3.9% and 3% respectively.
- *The major sectors for exports are*: Readymade Garments, Cotton Textiles, Textiles made from Man-Made fibre, Wool and Woolen Goods, Silk, Handicrafts, Coir and Jute.
- Exports in 2004-05: US $ 14 billion
- Exports in 2005-06: US $ 17 billion and registered 22% growth.
- Textile exports is increased by 11.43% in rupee terms.
- After the expiry of Multi Fibre Agreement in 1st January 2005, The share of Indian Textiles exports in imports of USA is increased from 4% to 5%.
- The share of Indian Textiles exports in imports of EU is increased from 6% to 7%.
- During 2005-06, India's total apparel exports were 8.63 billion US $.

Category	2002-03	2003-04	2004-05	2005-06	
Cotton Textiles	3.62	3.68	3.54	4.49	Manmade Textiles
1.53	1.86	2.05	2.00		
Silk	0.49	0.56	0.59	0.69	
Wool	0.29	0.35	0.42	0.47	
Ready Made Garments	5.75	5.92	6.02	7.75	
Handicrafts	1.42	1.11	1.01	1.24	
Jute	0.20	0.25	0.28	0.29	
Coir and Coir	0.08	0.08	0.11	0.13	
Total	13.37	13.80	14.03	17.08	

Market Share of Major Countries in Global Textile Industry

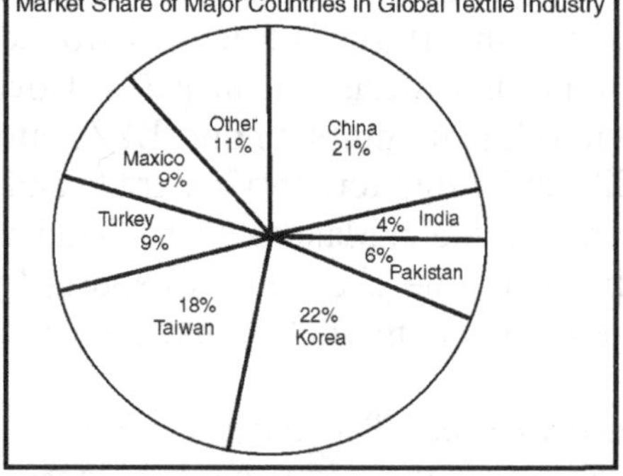

IMPORT LICENSING:

India has liberalized its Import regime for Textiles and apparel, but some of the part is still limited for market access. Currently, there is no import restriction for yarns and fabrics items. Apparel and Made-up textiles goods require a Special Import License (SIL). Govt. revised Exim Policy on 31st March 1999 by eliminating Import Licensing Requirements for 894 consumer goods, agriculture products and textiles. On 28th December 1999 India and Us signed an Agreement for the elimination of import restrictions of 1,429 agriculture, textiles, consumer goods and apparel. India removed restrictions on 715 tariff items as of 1st April 2000.

CUSTOM PROCEDURES:

Marking, Labeling, and Packaging Requirements: Marking, Labeling, and Packaging Requirements for Textile products are technically complex and difficult to implement. According to textile regulation passed on 22nd july 1998 by GOI, Yarns, and Fabrics to have the statutory markings and these markings should not mislead the consumers. For instance, Cloths must be remarked with the name and address of manufacturer, a description of cloth, sort number, length in meters and width in centimeters, and washing instructions.

The Man made fibre cloth must indicate whether it is made by spun or filament yarn. The month and year of packaging, the exact composition of cloth. The Marking must appear on the face plait of each piece of cloth. The language for marking must be in Hindi and English with international numerals.

EXIM POLICIES:

- *Duty Entitlement Passbook Scheme*: DEPS is available for Indian Export Companies and Traders on a Pre-Export and Post-Export basis. Pre-Export credit requires the beneficiary firm has exported during the preceding 3-year period. The Post-Export credit is a transferable credit that exporters of finished goods can use to pay or offset custom duties on imports of any unrestricted goods.
- *Export Promotion Capital Goods Scheme*: This scheme is available to export companies and traders who provide the GOI with information about which type of goods and what value of Capital Goods they will import. And they also inform what will be the outcome of export they expect to produce from those imports. Depending upon the export commitment GOI provides them a license to import capital goods duty-free or preferential rates of duty.
- *Pre and Post Shipment Financing*: The Reserve Bank of India provides Indian Exporters Pre-Shipment Financing through commercial banks for purchasing raw materials and packaging materials by presenting Letter of Credit. RBI also0 provides Post-Shipment Financing through commercial banks at preferential rates by presenting export documents.
- *Export and Special Economic Zones*: Govt. of India has established Export Processing Zones (EPZs) and Special Economic Zones (SEZs). In EPZs units can import goods free of custom duty. There is 5-year tax holiday to any industrial unit in EPZs. Govt. has allowed 100% Foreign ownership of units under EPZs and SEZs. The Govt. considers SEZs as foreign territory for trade and tariff purpose. Units

under SEZs may engage in Manufacturing, Trading and Services. Units are exempt from routine checking of exports by customs, and they can sell in the domestic market on payment of duty as applicable to imported goods.

Duty Drawback Scheme: The basic objective of this scheme is to reduce the indirect taxes on exports. Exporters can get refund of the excise and import duty. Through this scheme they can be more competitive and have more potential market.

TEXTILE INDUSTRY STRUCTURE

- The Indian Textile industry is highly fragmented sector.
- Industry is fully vertically integrated across the whole value chain and interconnected with various operations.
- Textile Industry comprises small-scale, medium-scale, large-scale, non-integrated, spinning, weaving, finishing, and apparel-making firms and enterprises.
- This is an unorganized sector and includes Handlooms, Powerloom, Hosiery, Knitting, Readymade Garments, Khadi, Carpet and Handicrafts manufacturing units.
- The organized Mill Sector comprises of spinning Mills, and Composite Mills where spinning, weaving, and processing activities are done.
- The Fibre and Yarn Sector of the textile industry includes Textile Fibres, Natural Fibres such as Cotton, Jute, Silk and Wool; Synthetic/Man-Made fibres such as Polyester, Viscose, Nylon, Acrylic and Polypropylene.
- The Man-Made Textile Sector includes Fibre and Filament Yarn manufacturing units of Cellulosic and Non-Cellulosic origin. The Cellulosic Fibre/yarn Industry is controlled by the Ministry of Textiles, and the Non-Cellulosic Industry is controlled by the Ministry of Chemicals and Fertilizers.
- India is the largest producer of Jute, the 2nd largest producer of Silk, the 3rd largest producer of Cotton and Cellulosic Fibre/Yarn and 5th largest producer of Synthetic Fibres/Yarn.

TEXTILES ORGANIZATIONS

There are several organizations and associations which are working for the smooth running of commerce of textile in India. They look after the welfare of the people involved in the production and delivery of the textile products. Watch out for some of such governmental, semi-governmental and private bodies.

ADVISORY BODIES:

- All India Handlooms Board
- All India Handicrafts Board

- All India Powerlooms Board
- Advisory Committee under Handlooms Reservation of Articles for Production
- Co-ordination Council of Textiles Research Association
- Cotton Advisory Board
- Jute Advisory Board
- Development Council for Textiles Industry

EXPORT PROMOTION COUNCILS:

- Apparel Export Promotion Council, New Delhi
- Carpet Export Promotion Council, New Delhi
- Cotton Textiles Export Promotion Council, Mumbai
- Export Promotion Council for Handicrafts, New Delhi
- Handloom Export Promotion Council, Chennai
- Indian Silk Export Promotion Council, Mumbai
- Powerloom Development and Export Promotion Council, Mumbai
- Synthetic and Rayon Textiles Export Promotion Council, Mumbai
- Wool and Woolen Export Promotion Council, New Delhi.

AUTONOMOUS BODIES:

- Central Wool Development Board, Jodhpur
- National Institute of Fashion Technology, New Delhi
- National Centre for Jute Diversification.

STATUTORY BODIES:

- Central Silk Board, Bangalore
- Jute Manufactures Development Council, Kolkata
- Textiles Committee, Mumbai

PUBLIC SECTOR UNDERTAKINGS:

- Central Cottage Industries Corporation Ltd., New Delhi
- Cotton Corporation of India Ltd., Mumbai
- National Textiles Corporation Ltd.
- National Jute Manufactures Corporation, Kolkata
- Handicrafts and Handlooms Exports Corporation of India Ltd., New Delhi

TEXTILES RESEARCH ASSOCIATION:

- Ahmedabad Textiles Industry's Research Association
- Bombay Textiles Research Association, Mumbai
- Indian Jute Industries Research association, Kolkata
- Man-made Textiles Research Association, Surat

- Synthetic and art silk –Mills Research Association, Mumbai
- Wool Research Association, Thane
- Northern India Textiles Research Association, Gaziabad
- South India Textiles Research Association, Coimbatore

INDIA TEXTILE POLICY 2000

For the growth and development of Indian Textile Industry and to make it more vibrant, Govt. of India passed National Textile Policy in 2000.

OBJECTIVES OF POLICY

- To produce and provide good quality cloth in affordable price to fulfill different needs of customers.
- To increase the share of India in Global Textile Market.
- To increase the contribution for employment and economic growth of country.
- Facilitate the Textile Industry to attain and sustain a pre-eminent global standing in the manufacture and export of clothing.
- Liberalization of controls and regulations for the market development of different Textile Segments and to make them stronger to perform in competitive environment.
- Encourage FDI and RandD to improve the manufacturing capabilities and infrastructure under the environmental standards.
- Facilitating financial support and arrangement to sector.

HIGHLIGHTS OF POLICY

- Achieving the target of Textile and Apparel Exports from US $ 11 billion to US $ 50 billion by 2010. The share of garments is targeted US $ 25 billion.
- Increasing contribution of private sector through set up environment-friendly and technologically advance textile units and complexes.
- To De-reserve the Garment Industry from Small Scale Industry.
- To improve the availability, productivity and quality of Raw Materials.
- Improvement in quality of fibre/Yarn and its availability.
- To increase the Cotton productivity by at east 50% and quality to match the international standard by implementing Technology Mission on Cotton.
- To make Silk and Sericulture sector strong, Govt. is aimed to improve R & D, Productivity of non-mulberry verities of silk.
- Encourage Private Sector to increase productivity, and to market the Wool products. Establishing Pre-Loom and Post-Loom facility.

- Developing high yielding seeds, to make farmers more technological efficient, creating potential market link, and diversified jute products are main concern for Jute Sector.
- Modernization and liberalization of Spinning Sector.
- Technological Up-gradation, Strategic Alliance with international textile majors, Improved Quality, and Productive HR environment are the concern for Organized Mill Industry.
- For Powerloom Sector steps such as, Technological Up-gradation, Increased Productivity and Modernization of the sector are going to be implemented.
- To Increase Productivity, Enhanced Marketing and Distribution, Technological Up-gradation and increased Research Development in Handloom Sector.
- To Improve Export, Promotion, Technological Support, Infrastructure, Marketing and Distribution and R & D for the growth of Handicraft Sector.
- To increase Exports to 50 billion US $ by 2010, Increase FDI and Investment, and to Increase Infrastructure Support to export oriented areas.

TEXTILE POWERLOOM SECTOR

This part of industry includes fibre and filament yarn manufacturing units. The production of man-made fibre has shown promising growth over the years. The total production of fibre is expected to increased by 14% during year 2006-07. The production of Polypropylene staple fibre and Polyester Fibre is expected to increase by 26% and 11% respectively. The production of Polypropylene Filament Yarn and Polyester Filament Yarn is expected to increase by 12% during 2006-07

THE POWERLOOMS SECTOR

The Powerlooms sector is decentralized and plays vital role in Indian Textiles Industry. It produces large variety of cloths to fulfill different needs of the market. It is the largest manufacturer of fabric and produces a wide variety of cloth, both grey as well as processed. There are 19.54 lakh powerlooms in the country distributed over 430,000 units. The sector contributes around 62% of the total cloth production in the country and provides ample employment opportunities to 4.86 million people.

POLICY BRIEF ON HANDLOOM SECTOR 2010

Handloom sector is a very ancient industry in India. Features of this sector vary across India. In some parts, it has attained the status of a mature industry, and in other parts, it is still an enterprise confined to the needs of the particular home. Given the changes in the society in the last 100 years, there has been growth of mechanized textile production across the world, principally driven by Industrial revolution and by Western countries. Gradually, due to competition and for other reasons, handloom has lost much of its markets, and is almost non-existent in most countries. However, handloom sector is still a force to reckon with in India, and some of the South Asian and Asian countries such as Srilanka, Bangladesh, Thailand, and Cambodia. In India, textile sector today comprises four important segments viz., modern textile mills, independent powerlooms, handlooms and the garment sector. Though there is a huge, informal tailoring enterprise it has never been studied or considered for any kind of attention, leave alone policy.

There are also sub-sectors like textile machinery manufacturing and spinning sectors, which form part of the textile sector. Among all these, handloom sector being the oldest has different features in different parts of the country. It employs the largest number of people. However, it is considered as a sunset industry, and there is an air of inevitability given the relentless march of mechanization, modernization and sophistication. However, there are many advocates among the Indian population, who support the cause of handlooms for various reasons, including ideology, philosophy, sheer love for handloom products, economic arguments, etc... Irrespective of the policies, projects, and aspirations arising out of various quarters, handloom sector is undergoing changes which are impacting on the livelihoods of the handloom weavers. Some of the factors are internal and some are external factors related to the growth of modern textile sector. Before discussing these changes, it would be worthwhile to note some of the hard facts (substantiated by statistics) of the handloom sector.

FEATURES OF HANDLOOM SECTOR

Handloom sector is an important cottage industry in India, and is a very old profession. Handloom weavers are known for their knowledge, innovation and brilliance in designs. Weaving is now considered almost an art from, considering deployment of skills and knowledge.

This sector, estimatedly, provides employment for more than fifteen million people and is second largest rural employment provider next to agriculture. In Andhra Pradesh, there are more than 3 lakh handloom weaving families. In comparison with other traditional rural sectors, handloom weaving is a full-time family profession, involving all the members of the family.

In Andhra Pradesh, there are about 3,20,000 handlooms. About 5,00,000 families are directly, while indirectly 20,00,000 families, dependent on them. Major handloom centers include Chirala, Mangalagiri, Pedana, Ponduru, Polavaram, Pochamapalli, Puttapaka, Gadwal, Dharamvaram, Emmiganur, Madhavaram, and Narayanpet.

Each center has its unique identity of producing varieties of handloom products. Large numbers of handloom can be seen in north coastal Andhra (Srikakulam, West and East Godavari district), south coastal Andhra (Krishna, Guntur, Prakasham, Nellore), Rayalseema (Cuddapah, Kurnool, Chittoor, Anantapur) and Telangana (Nalgonda and Warangal district). In districts such as Karimnagar, Medak and Vijayanagaram, a few thousand handlooms can be seen. In all centers of Andhra Pradesh, production is geared for domestic market only.

In the entire country, there are more than 38,00,000 handlooms. However, in north-eastern States, there are more than 15,00,000 domestic handlooms. Handlooms in north India and South India are geared for commercial production catering to the needs of domestic market and also exporting their products abroad.

Estimatedly, there are about 12.5 million people dependent on handloom weaving. In view of this, the national and State governments have provided for schemes to safeguard the interests of the weaving community. These schemes pertain to production inputs, market support, development and welfare of the weavers. However, in spite of the legislative policy safeguards and supportive schemes involving huge budgeting, financial allocations, the handloom industry has been experiencing periodic crises. The principal cause being the bias towards modern technologies and industrialization, and inadequate understanding of the socio-economic realities relevant to the textile sector.

Formal education system has not included this important profession in its fold. As a result, any innovation and change has been left to the handloom weaving families. Over years in the recent decades, due to lack of information and fast paced changes, practices in handloom sector became static and apparently became redundant. Formal research institutions, Universities and other institutions have not included this sector in their working agendas.

In the era of globalisation, information access has become very critical, more so for traditional communities who are dependent on professions that have extreme relevance as livelihood platforms for crores of people. Handloom method of cloth production is also environmentfriendly, and would be supportive of sustainable development policies aimed at reduction of negative impacts on environment and ecology. Handloom sector is an important channel for balanced sustainable economic growth.

Presently, handloom weavers are facing severe livelihood crisis because of adverse government policies, globalisation and changing socio-economic conditions. While earlier advocates of handloom sector within policy making circles were extremely popular, and their arguments were valued, presently there is no longer a protagonist section. As a result, decisions of the government have been extremely adverse, retrograde and detrimental to the interests of the handloom weavers.

In this information age, information availability and access, in different forms, is crucial for growth and also enhances the ability of the sector to remain competitive and vibrant. Textile sector in India is growing, and has been undergoing enormous changes within its structure, which are affecting its characteristics. The profession of handloom weaving has been one of the main factors of growth in Indian economy, promoting inter- and intra-relationship between different communities. This sector has been considered important for the sheer magnitude of people endowed with traditional artisan craft skills attuned to the local needs and resources.

The scope and range of the sector is now being enhanced by the growing relevance of environment-friendly technologies in the clothing industry. Awareness and debate from local to global level on these technologies is being promoted by the policies of WTO and several environmental organizations in their given spheres.

However, in the present situation, there are too many issues, which are impinging on the development of this sector. There is a need to develop broader understanding of its multidisciplinary perspective, which is more needed in the present circumstances of globalisation and environmental degradation, particularly in relation to development.

Presently, government policies are increasingly influenced by the globalisation processes and are related to WTO-induced trade regimes. As the controls on exports get liberalized and the domestic markets become open to imports, the textile scenario in the country is likely to undergo drastic changes in terms of skills, inputs like designs, market trends and changing demands therein.

The question before us is: whether these changes include the interests of large number of weavers who are still practicing their skills and knowledge, honed over centuries? Do we have a moral obligation to preserve certain amount of these skills to the coming generations? If so, where do we strike the balance between mass production and production by the masses? Is there a need to work out viable models of sustainable utilization of resources?

Ineffective implementation of the schemes meant for handloom development and the changed context of textile industry - increasing competition from the powerloom and mill sectors – has been responsible for the crisis in the handlooms. However, lack of information to the weavers regarding various policies and schemes is no less a significant cause for the dwindling fortunes of the weaver community.

In addition, there appears to be an inadequate appreciation among the legislators and policy makers of the actual state of the handloom sector, the appalling conditions of weavers and the fate of the policies meant for them, and the consequences therein for the individual, communities and the economy. It is saddening to note that even the government departments and implementing agencies related to handlooms suffer from inadequate information and data resulting in a widening gap between policy formulation and implementation.

Handloom sector was a nationalist activity and identified completely with Gandhian agenda. Handloom weaving and related activities became symbolic for the Indian Independence struggle.

In earlier planning at the national level, development of handloom sector was seen as stimulation for rural development, considering the usage of local resources, local craftsmanship and catering primarily to local markets. In the first decades of India's Independence, all national policies emphasized on this. However, now the thinking at the apex policy level is that handloom sector is a redundant profession and is a burden on the government exchequer.

However, for somebody at the ground level, it is clear that the present policy discourse is devoid of the realities of the handloom sector, and has generally avoided inputs from the handloom community. Political leadership, in general, has been avoiding taking up the cudgels on behalf of this community. Consequently, there is no discourse on handloom policies and also the replacement/downsizing

of the whole set of existing policies. Handloom weavers have no say in policy matters. They have no channels of communication and information. Despite various welfare policies, handloom weavers are poor and vulnerable and their economic situation is precarious. Corruption is pervasive. Handloom weavers have lost faith in these structures. However, their choice is limited.

STRENGTHS OF HANDLOOM SECTOR

Handloom sector continues to employ large number of people in rural, semi-urban and urban areas of India. In some states such as Andhra Pradesh, Tamilnadu, Kerala and Karnataka, Uttar Pradesh, Gujarat, Rajasthan, North Eastern States, Madhya Pradesh, Orissa, West Bengal, this sector is visibly large and dominant in certain categories of clothing. Market for handloom products is still large and wide. There is good domestic market as well as international market. There are die-hard consumers who would 'support' handloom products for every reason they can hold onto.

Handloom sector continues to provide direct and indirect employment. There are estimatedly 32 other sectors which are dependent on handloom production, in various ways, including transportation, financial services, marketing services, service and maintenance services, hotels, etc. Many handloom centres are well known tourist spots, drawing visitors from far places of India and foreign countries as well. Thus, part of the tourism industry's fortunes is also influenced by handloom sector and its fame.

Handloom sector has umbilical linkage with cotton farmers and rural farm economy. Agricultural labour gets employment in handloom sector in non-agricultural seasons. Handloom sector has self-sustaining mechanisms, including training for young weavers, irrespective of gender. The inheritance of skills, resources and capacities is beyond the realm and reach of any modern training and educational institution.

It is a facilitation process, which is not dependent on the government and or any modern formal institution. There is also sufficient flexibility for all types of communities to take up handloom production as a profession. Handloom sector is part of the culture and ethos of India and its glorious past. It has emotional bondage with nationalism and the champions of nationhood. It had a principal role in public opinion formation during the Independence struggle against the 'imported' goods and 'imposed' industrialisation.

Handloom production has significant contribution to the national GDP and export earnings. Thus, it has some influence over the foreign exchange levels and the well-being of the economy.

RELEVANCE IN MODERN ECONOMY

Handloom sector has different strengths in the modern economy. It is simple, appropriate technology, the knowledge of which lies with the people. Knowledge dissemination is not locked in any consultancies, or training institutions. The very essence of this technology is that it can be assembled by anybody within no time.

It is not expensive either to erect or dismantle it. In a world, which is gasping for clean air and water, and vibrant soil, caused by the pollutants and impacts of the modern textile mills and powerlooms, handlooms are eco-friendly. The environmental impact of a handloom is far below its more modern competitors. It is an independent and autonomous technology. Energy impacts from handloom technology are almost zero.

While the world is becoming modern and globalisation is happening, fashions are also changing rapidly. Fashions are increasingly becoming customised-production *i.e.*, productions of clothes or garments for particular person. This is possible on handloom with much more ease and lesser cost. Customised textile production is possible only with handloom technologies.

CONTINUING CRISIS

Handloom sector has been in crisis in different time for different periods, for the same set of reasons. Earlier attempts by the governments in the 1960s did address some of these issues, though they could not solve them completely. During the ATC-scenario, between 1995-2005, given the apathy and discrimination of the government, the very same issues with new set of factors have once again pushed handloom weavers into crisis. Often, in discussions and deliberations on handloom sector, symptoms are recognized as factors.

As a result the whole discussion is becomes cyclical, with no conclusion reached. It is important that critical factors are addressed, and not the symptoms or consequences of critical factors. Further, it is important to come up with a suggestion that addresses the critical factor, and not the symptom alone. Any industry, the world over, requires public policy support. Handloom sector is no exception.

While it is true that handloom sector is major source of employment, one cannot ignore the fact that it is also an industry. The strength of the handloom sector is its large and skilled labour. This is also the strength of the Indian textile sector. One needs to enhance this strength and address factors that weaken this strength. Modern economists should keep in mind the fundamental economic principle: maximum good for the maximum people.

CRITICAL FACTORS

There are many problems in handloom sector. Most of the factors which cause these problems have their origin in conditions, which are outside the control and access of average handloom weaver. This viewpoint is at variance with the popular notion in the textile policy circles that problems in handloom sector start and end with the attitudes of the weavers towards change, and the social change towards mechanized production.

Market demand for handloom products establish the fact that handloom has survived the consumer attitude. Hence, the following problems of handloom sector have to be viewed from point of policy interference, like any other production sector, which has been contributing to GDP and national growth.

Raw Material Supply

Access to raw material such as yarn, dyes and dye stuffs has become a problem. Since this is a rural and semi-rural production activity, weavers have to go far to get these raw materials. On the top of it, the prices are steadily increasing. Hank yarn availability is serious issue because it is controlled by modern spinning mills, who see more profit large-volume cone yarn than hank yarn. Secondly, since hank yarn is tax-free and has subsidies, enormous amounts are diverted to powerloom and mill sector. As a result, there are perennial shortages of yarn. Despite a few schemes, hank yarn access has not been solved, and the seriousness continues to grow. Colours have been expensive, and presently there is no system or mechanism to increase their availability to weavers.

The Hank Yarn Obligation was introduced in 1974 to require spinning mills to process 50 per cent of their deliveries in hank form, with at most 85 per cent in 40s count or below. However, this Obligation did not solve the problem of access to yarn for the handloom weavers for various reasons. The Hank Yarn Obligation (HYO) requires spinning mills in practice either to process about a quarter of their deliveries in the form of a yarn used by handlooms or to transfer the obligation to other firms. Manually reeled, hank yarn is rarely made in-house since production uses oldermachinery and requires a work-force (mostly women) half as large as that of the total factory.

Instead, most mills pay others from 25 paise to Rs 1 per kilo to fulfill the obligation, itself calculated on the basis of a percentage of the net qualifying amount after exemptions for exports, for own consumption in composite mills, and for hosiery and blended yarn output.

Tax exemptions on hank yarn, for instance, make it profitable to declare cone yarn (used by powerlooms and composite mills) illegally as hank yarn. Since the combined effect of tax exemptions and production controls is to lower hank yarn prices relative to cone yarn, non-handloom weavers often use the former. The leakage is estimated at 15-40 per cent. The Planning Commission Working Group on Textiles for the 11th Five Year Plan recommended that the Hank Yarn Obligation Scheme may be reviewed, and the possibility of evolving an alternate mechanism to ensure the supply of adequate quantities of yarn to the handloom sector may be considered.

Government of India introduced the Mill Gate Price Scheme in 1992-93 with the objective of providing all types of yarns to handloom weavers' organizations at the mill gate price. The National Handloom Development Corporation (NHDC) is the only agency authorized to Implement the scheme. NHDC has got 110 functioning Yarn Depots set up by associating

State Handloom Corporations/Apex Bodies/Weavers Cooperatives/ Exporters/Master Weavers, etc., from where yarn is being supplied to weavers at mill gate prices. Additional 273 yarn depots have been opened this year. However, this scheme is miniscule, compared with the need in terms of quantity, quality and timely availability. With the market demand growing to Rs.1,00,000 crore, handloom sector requires more and more allocation for such a scheme. Year wise supplies of yarn in quantity and value terms under the Scheme during Tenth five year plan are as follows:

Table. Supplies of Yarn during the Tenth Five Year Plan

Year Qty (in Lakh Kg)	Value (in Rs. crore)
2002-03	397.98 331.11
2003-04	242.77 240.71
2004-05	202.38 223.08
2005-06	220.86 228.16
2006-07	186.67 171.77
(Upto 20th Oct., 2006)	

Market for Products

Regulations were imposed on both mills and powerlooms to protect the handlooms from their unequal competition. The most important among these was the reservation of products for the exclusive production of handlooms. However, the reservation introduced for 8 items in 1950 was not very effective. Out of these 8, 5 items were reserved for both handlooms and small powerloom units. To correct this, the Handloom Reservation Act of 1985, reserving 22 items for the exclusive production of handlooms was introduced. However, it could not be implemented till 1993 due to the legal dispute posed by the mill and powerloom lobby in the Supreme Court. To overcome this, Abid Hussain Committee recommended the inclusion of the Act under the Ninth Schedule of Constitution but this was never considered (GOI: 1990).

The Mira Seth Committee recommended the reduction of the number of items reserved from 22 to 11 and recent annual report of Ministry of Textiles indicates that this has been implemented (Ministry of Textiles: 2001).

Even after the Act came into effect from 1993, the number of violations reported in the annual reports has been shocking low, to the tune of five to fifteen from all over the country (Ministry of Textiles: 2001). Thus, despite the demands of weavers the

Reservation Act has never been implemented efficiently and effectively. In addition to this, the approach paper to Tenth Plan [Approach Paper to Tenth Plan (2002-7): p19], suggests that reservation is uneconomical in the wake of liberalization and hence has to be phased out eventually. Most State governments including the GOI did not dare to expend the budgets earmarked for enforcement.

In 2001-02, AP government did not expend anything on enforcement despite availability of funds. Accompanying Table demonstrates the lax attitude. For example, it is widely known that Uttar Pradesh (Mau and Benares), from where Kota Doria sarees produced on powerlooms are being exported to Kota in Rajasthan, as handloom products where there are no machinery.

Enforcement allocations to Rajasthan does not make any sense in this case. Similarly Narayanpet sarees, of Andhra Pradesh, produced on powerlooms in neighbouring Karnataka, do not get any protection. It is not just aninter-State problem but intrastate enforcement across the country is lax.

This does not receive the required attention by government officials as they tend to brush this away as irrelevant in their work. Statement Showing Assistance Released under Implementation of the Handlooms Act, 1985 for setting up of the Enforcement Machinery (Rs. in Lakhs) as on 29/08/2005.

S. No.	**Name of the State**	**Amount Released during Xth Plan 2002-03**	**2003-04**	**2004-05**	**2005-06 (Apr-Aug.)**
1	Andhra Pradesh	17.93	0.00	0.00	-
2	Bihar	0.00	0.00	0.00	-
3	Gujarat	18.66	39.25	19.20	34.18
4	Haryana	6.05	7.68	5.60	-
5	Kerala	0.00	0.00	0.00	-
6	Madhya Pradesh	6.88	5.99	7.19	-
7	Rajasthan	8.73	10.11	23.41	4.19
8	Tamilnadu	61.75	38.08	64.60	41.16
9	West Bengal	0.00	18.89	0.00	19.66
Total		120.00	120.00	120.00	99.19

The Working Group of Planning commission for 11th Five year Plan recommended the review of this. Handloom products require more visibility. This means better and wider market network. One-off exhibitions organized with the support of government might support, but do not suffice. Presently, handloom products are available only in fewer places.

Working Capital

The entire spectrum of policy interventions for credit delivery is centered around cooperative institutions consisting of weavers cooperatives as recipients of credit and the cooperative banking system as providers of credit services1. Here, the National Bank for Agriculture and Rural Development (NABARD) plays the crucial policy making role.

NABARD estimates that only 20 to 25% weavers operate within the cooperative fold and that this number is also on the decline. This implies that, by the very nature of the intervention, over 75% of the weavers today are left out of the formal financial delivery mechanism. As a result of institutional preference, there were no significant policies in the past to provide credit to weavers and stakeholders outside the cooperative sector. Only recently has NABARD started to respond to other stakeholders.

Over the last seven years, credit flows (utilisation) has declined from Rs 586 crores in 1998-99 to a level of Rs163 crores in 2004-05. A worrisome feature of this decline has been the growing gap between the cash credit limits sanctioned by NABARD and the actual amounts utilised by the State Cooperative Banks (SCBs).

In 1998-99, the overall utilisation by SCBs was 78% of the sanctioned limits, but this gap increased to 47% in 2004-2005, clearly indicating that in spite of higher available CC limits, the cooperative system is not able to utilise it. In one district cooperative bank, of the sanctioned credit limit of over Rs 3 crores to the Weavers Cooperative Societies (WCSs) for the current year, the withdrawals were just Rs 5 lakhs. There are WCSs who have even stopped using CC limits.

One of the important reasons for the failure of the formal institutional credit delivery mechanism has been the precarious financial position of the District Central Cooperative Banks (DCCBs) and the SCBs entrusted to extend credit facilities to the WCSs. While the number of SCBs incurring losses is only 4 against the remaining 27 that are profit-making, for the DCCBs, the situation is not the same.

Of the 365 DCCBs in the country, in 2003-2004, 102 were in losses and remaining 263 in profit. The losses were Rs 758 crores. According to NABARD data, accumulated interest and total loans outstanding with all the WCSs and their apex cooperative structures of 30 states is Rs 1,008 crores at the end of 2004-2005.

The Artisan Credit Cards (ACC) introduced in 2003 offer a potential credit delivery channel to reach a large number of weavers within and outside the cooperative fold. In principle, its design also offers the potential to reach weavers who have been stuck with a non-performing cooperative structure at the village level.

NABARD in 2005 has introduced guidelines for banks to experiment with Self-Help Groups (SHGs) of weavers. The experience in SHG promotion methodology shows that it takes at least one year for the group to develop a healthy group dynamic in addition to regular savings. The group then becomes credit worthy.

POLICY TURNS AND FAILURES

Handloom sector is amidst a crisis borne out of policy apathy and bad governance. While the markets have been responding by increasing the demand for handloom products, handloom weavers are finding it harder and harder to cope with low wages, markets risks and rising cost of production. There are number of factors which led to this situation.

Appallingly, the response of the government has been poor and negative. While every policy document talks about the need for growth and sustenance of employment, handloom sector does not deserve the attention at a level that is required.

Broadly, three phases can be seen for handloom sector in Five-year Plans:

1. *Early phase included*:
   - Increasing employment
   - Increasing production
   - Cooperativisation
   - Protection
   - Ensure supply of inputs
2. *Second phase included*:
   - Modernisation
   - Protection
   - Welfare
3. *Third Phase includes*:
   - Modernisation
   - Competition
   - 'Market is the answer' approach

PRE-1985 PERIOD

For a long period, after Independence, government was pushing for modernization in Indian textile industry. This period, however, government through Five year plans was committed to the growth of handloom sector. Handloom sector found place in all Five year Plans. In the 1960s and 1970s, there was huge thrust on the formation and encouragement of handloom cooperatives. These cooperatives increased the access of weavers to both inputs and markets.

1985-1995 PERIOD

In 1985 government announced a new textile policy. This was a period when Mr. Rajiv Gandhi wanted to take India into 21st century - meaning to modernize. Modernisation became the slogan of the government, and the ruling clique. Soon, textile policy and governance started reflecting this trend. The New Textile Policy for the first time acknowledges the need to diversify into man-made fibre-based textile production.

It also acknowledged the presence of growing number of powerlooms piggy backing on government support. At this stage, a slogan was floated, "Cotton for the masses and polyester for the classes". Reliance and its brand (Garden Vareli) established their market presence in this period. This was done to placate handloom sector and its champions who were still powerful.

A Handloom Reservation Act was also brought confining handloom products to 22. All other textile products were open for competitive production and marketing. But this Act was challenged, and after so many years, Supreme Court had upheld the Act. By then, government had lost its interest in the implementation of the Act.

With cotton farmers committing suicides in 1987 in AP, and in subsequent years by the handloom weavers, there was a notice of this new phenomenon. But this did not deter the government to push for economic reforms in early 1990s, through the combine of Dr. Manmohan Singh and Sri. P. V. Narasimha Rao.

As India started moving into WTO regime, despite protests, there were not many voices asking for the cause of handloom weavers. Meera Seth Committee in 1995 recommended the downsizing of the reserved handloom products from 22 to 11.

1995-2000 PERIOD

With India signing the WTO agreement, which included 28 agreements, India has indicated its agreement to compete in textile trade at the global level, by freeing its own markets. One of these agreements included Agreement on Textile Clothing. While everyone was worried about surge in imports, not many noticed other developments in Indian textile sector.

In 1997, World Bank charted a positive growth scenario for Indian textile industry through exports. An export-oriented policy was clearly asked for. However, for this to happen, World Bank recommended the removal of restrictions such as Handloom Reservation Act, Hank Yarn Obligation on spinning mills and withdrawal of support to handloom sector. It wanted India to modernize its Indian textile industry in anticipation of opening of textile markets in US and Europe.

In 1999, Satyam Committee taking cue from the World Bank and other neo-liberal writings, recommended the same. However, following popular protests, government declared in the Parliament hat these recommendations would be put on hold.

However, with the announcement of new textile policy 20002, it was clear that the government has set its mind on the course charted by World Bank and Satyam Committee. Thus, the central government and State governments started reducing their attention to handloom sector by slowly scorching the cooperative sector, which was the mainplank of handloom sector's competitiveness.

2000-2010 PERIOD

Budget allocations for handloom sector remained the same, if not registering a marginal growth. While allocations for non-handloom textile sector was growing phenomenally, handloom budgets declined in real terms4. Cooperative sector became weaker. Handloom weavers suffered from low wages and diminishing access to inputs and markets.

Government did not bother to implement any of the Acts and rules meant to protect handloom weavers. As a result, powerloom products started entering handloom markets, as 'imitation' products. This has further eroded the handloom markets and thus impacted on the production of handloom production. Over a period, this made handloom sector more and more vulnerable to slightest shift in prices of inputs and products in the markets.

Now this is being used by the government and others that handloom sector in incompetitive. The draft National Fibre Policy5, as developed, is poised to give another blow to the handloom sector. This policy recommends encouragement of man-made fibres, at the cost of natural fibres. This would increase the costs of natural-fibre based textile products, in comparison with man-made fibre products, essentially killing all the small-scale production.

This policy would at a stroke increase the impact to 10 crore population dependent on natural fibre based textile supply chains. It would change the cost of production and consumption patterns of textiles. It would influence the textile markets and increase scope for growth of manmade fibre textile products. It would reverse the slogan given earlier, "Cotton for the Classes and Polyester for the Masses"

CONCLUSION

Often, in discussions and deliberations on handloom sector, symptoms are recognized as factors. As a result the whole discussion is becomes cyclical, with no conclusion reached. It is important that critical factors are addressed, and not the symptoms or consequences of critical factors. Further, it is important to come up with a suggestion that addresses the critical factor, and not the symptom alone. Any industry, the world over, requires public policy support. Handloom sector is no exception.

While it is true that handloom sector is major source of employment, one cannot ignore the fact that it is also an industry. The strength of the handloom sector is its large and skilled labour. This is also the strength of the Indian textile sector. One needs to enhance this strength and address factors that weaken this strength. Modern economists should keep in mind the fundamental economic principle: maximum good for the maximum people.

It is time that government recognized the value of the handloom sector in achieving sustainable development of the country. On its own, government would never be able to provide employment to such a large workforce. Going by the logic of liberalization, government in turn ought to formulate, promote and encourage policies which sustain this employment, and cannot work to its detriment.

Despite adverse conditions, due to larger support from the consumers and being a livelihood option for millions of weavers, handloom sector has been surviving, and has the potential to be so6. Government has to ensure a 'level playing field' for this sector towards healthy competition among the different sub-sectors of textile industry. Presently, handloom weavers are increasingly finding it difficult to continue their traditional occupation, and are not capacitated to find any other job for living. Their love and skill is being tested every day, with growing inflation. About 3 crore people and their families are in crisis.

TEXTILES MAJOR COMPANIES

Following are some major players in the vast field of Indian Textile Industry.

ARVIND MILLS

Arvind Mills is one of the major and fully vertically integrated composite mills player in India. It has large production in denim, shirting and knitted garments. It is now adding value by manufacturing denim apparel. Its sales are around US$ 300 million.

RAYMONDS

Raymonds has the large, diversified integrated business model, which is spread across the value chain from yarn to retail. It is specialized in Diversified woolen textiles. It already supplies to some US retailers. It also looking to also expanding denim capacity and to become second largest denim player in India. Its presence in retail will be big positive in future. Its annual sales are around US$ 300 millions

RELIANCE TEXTILES

Reliance Textiles is one of the major Textile Company that is in business of fully integrated manmade fibre. It has capacity of more than 6 million tones per year. It has joint venture partners like, DuPont, Stone and Webster, Sinco (Italy) etc.

- Vardhaman Spinning Vardhman deals in spinning, weaving and processing segment of the industry. It is planning to double its fabric processing capacity to 50 million meters. It is an approved supplier to global retailers like GaP, Target and Tommy Hilfiger. Its sales are little over US$ 120 millions
- Welspun India (Manufactures terry towels)
- Century Textiles (Composite mill, cotton and Man-made)
- Morarjee Mills (Fully integrated Composite Mill)
- Indo Rama (Cotton and Man-made)
- GTN Textiles (Cotton Yarn and Knit Fabrics)
- Ginni Filaments Ltd. (Yarn and Fasbric)
- LNJ Bhilwara Group (Diversified and vertically integrated denim producer with spinning and weaving capacity)
- Mafatlal Textiles (Fully integrated Composite Mill)
- Modern Group (Diversified, producer of denim, syntax and thread)
- Ashima Syntex (Man-made Fibre)
- KG Denim (Fabrics)
- Sanghi Polyesters Ltd. (Manmade Fibre)
- Nova Petrochemicals (Man-made Fibre)
- S. Kumar Synfabs Ltd. (Home furnishing and Suit Fabrics)
- Bombay Dyeing Ltd. (Composite and fully integrated)
- Rajasthan Petro synthetics (Diversified)
- BSL Ltd. (Textiles)
- Garware Polyester (Diversified)
- Banswara Syntex (Composite)
- National Rayon Corp. (Man-made fibre)
- GSL India Ltd. (Threads)
- Indian Rayon (Man-Made Fibre)
- Alok Textiles (Cotton and Man-made Fibre Textiles)
- Sharda Textile Mills (Man-made Fibre)
- Birla Group Dormeuil Birla VXL Ltd. (Fully integrated woolen textiles)
- Gokuldas Images (Diversified)
- Hanil Era Textiles (Yarn, Cotton and Man-made Fibre)
- Oswal Knit India (Woolen Wear)
- Niryat Sam Apparels (Apparel)
- Filaments India Ltd. (Manmade Textiles)

BRIGHT FUTURE OF INDIAN TEXTILES

The textile industry in India is substantial, and largely diversified. It is hard to cover all its aspects even within hour-long presentation, however, research team of Fibre fashion.com has attempted to gather significant pieces of massively complicated puzzle of this industry. The Indian textile industry roots thousands of years back. After, the European industry insurrection, Indian textile sector also witnessed considerable development in industrial aspects. Textile industry plays an important role in the terms of revenue generation in Indian economy. The significance of the textile industry is also due to its contribution in the industrial production, employment. Currently, it is the second largest employment provider after agriculture and provides employment to more than 30mn people.

PROSPECT

Considering the continual capital investments in the textile industry, the Govt. of India may extend the Technology Upgradation Fund Scheme (TUFS) by the end of the 11th Five Year Plan (till 2011-2012), in order to support the industry. Indian textile industry is massively investing to meet the targeted output of $85bn by the end of 2010, aiming exports of $50bn.

There is huge development foreseen in Indian textile exports from the $17bn attained in 2005-06 to $50bn by 2009-10. The estimation for the exports in the current financial year is about $19bn. There is substantial potential in Indian exports of technical textiles and home-textiles, as most European companies want to set up facilities near-by the emerging markets, such as China and India.

The global demand for apparel and woven textiles is likely to grow by 25 per cent by year 2010 to over 35mn tons, and Asia will be responsible for 85 per cent output of this growth. The woven products output will also rise in Central and Southern American countries, however, at a reasonable speed. On the other hand, in major developed countries, the output of woven products will remain stable. Weaving process is conducted to make fabrics for a broad range of clothing assortment, including shirts, jeans, sportswear, skirts, dresses, protective clothing etc., and also used in non-apparel uses like technical, automotive, medical etc...

It is been forecasted that the woven textile and apparel markets will sustain their growth from current till 2010. The imports of apparel and textiles will rise from developed economies like the USA and the western countries of Europe and Japan, along with some newly emerged economies, such as South Korea and Taiwan. Certainly, import growth has been witnessed vertical rise in the previous year.

Apparel is the most preferred and important of all the other applications. Woven fabrics are widely used in apparel assortments, including innerwear, outerwear, nightwear and underwear, as well as in specialized apparels like protective clothing and sportswear. Home textile also contributes considerably in woven fabric in products assortments like curtains, furnishing fabrics, carpets, table cloths etc.

Special kind of woven fabrics are utilized in medical as well as industrial applications. The medical applications include adhesives, dressing bandages, plasters etc. Where as, industrial applications includes; o Geotextile - interior upholstery, trim, airbags and seat belts and lyre fabrics. Sailcloth - tent and fabrics used architectures, transportation and tarpaulins. And many more applications...

The Indian Industry foresees huge demand for industrial woven products for medical and automotive applications. Demand for woven fabrics is anticipated to be rise vertically in the sector of home textiles. Non woven sector has great future in terms of global demand, thus major facilities of cotton yarn are currently concentrating just on home textiles. It is mandatory, that the peak management of the cotton yarn manufacturers analyse the future prospect and growing graph of demand for non woven products.

CONCLUSION

Anticipating massive growth in medical and automobile sectors, these sectors assures substantial demand for non woven facilities in India. Albeit, home textiles also will lure higher demand, there are specific demands for home textile facilities also.The 7th Five Year Plan has huge consideration on agricultural growth that also includes cotton textile industry, resulting a prosperous future forecast for the textile industry in India.

Indian cotton yarn manufacturers should rush forward for joint ventures and integrated plans for establishing processing and weaving facilities in home textiles and technical textiles in order to meet export target of $50bn, and a total textile production of $85bn by 2009-2010.

INDIAN FABRIC MACHINERY INDUSTRY

India's industrialization in various other sectors has been successful due to the resources generated by the textile industry. The Indian textile industry is the second largest industry across the world. As a result Indian textile art is famous across the world. India is considered the third largest producer of cotton, as it has the largest cotton acreage of 9 million hectares.

India stands fourth in the production of the stable fibre and sixth in the production of the filament yarn. Nearly 30% of the India's export is from the textile industry that covers textiles and garments. Textile industry comprises of various segments such as cotton textile including handlooms, silk textiles, manmade textiles, woolen textile, readymade garments, handicrafts including carpets, jute, coir, and many more.

CURRENT STATUS OF THE TEXTILE MACHINERY INDUSTRY

Textile industry is the most essential necessities of the people. Textile industry accounts approximately 14% of the total industrial output. Also total export of the textile industry amounts to 30%. In terms of employment generation also this industry is the largest. This industry generates massive potential for employment from agricultural to industrial.

CURRENT DEVELOPMENT IN TECHNOLOGY

Textile machinery in India has improved a lot and there has been fast progress in technology used in the machinery also this has been beneficial for textile manufacturer in India. The textile traditions have still remained in some way or the other. In the last five decades the textile industry has developed a lot and played vital role in the transformation of this industry. Regular efforts are being made to scale up research and development. Also the ways have been looked that limits dependence on the foreign manufacturers in order to supply advanced technology.

IMPORTANT PLACES TO SOURCE TEXTILE MACHINERY FROM INDIA

The major hub of the textile machinery manufacturing industry in India is Coimbatore in Tamil Nadu. Major machines manufactured here include knitting and spinning machines, wrapping machines, textile testing equipments, and many more. Other important destinations in India include Mumbai, Ahmadabad, Surat, Ludhiana, Rajkot, Kolkata, Chennai, Bangalore, Panipat, Bhilwara, Kanpur, Gurgaon, Faridabd, Noida, Hyderabad, Gaziabad, Thane, Pune, Amritsar, and many more.

DEMAND OF THE INDIAN TEXTILE MACHINERY IN INTERNATIONAL MARKETS

Indian textile machinery manufacturers produce stylish machines that offer higher speeds as well as production capability. All this is offered at competitive prices. IT sector has provided strong support to this sector. The dependence of the imported machinery also reduced as the new technology developments led to the development of home grown machines. Various textile machinery have huge demands in the International markets. These include Crushing machine, Wrap Crushing Machine, Felt Calendar, Super Embroidery Decca Machine, Baby Boiler Machine, Standing Pleating, Hydraulic Embossing Machine, and many more.

4

Marketing Trends in Textile Industry

A high technology industry will employ fewer employees, but these will be more highly educated. The postulation for management and technology trained people will increase, but clerical jobs will decrease. The causes for the decline in employment are: import penetration, higher productivity with new technology, and the death of appropriate marketing aggressiveness of the manufacturers. These factors influence the different industry segments in various degrees. In the aggregate, rising imports influence the growth of the industry but not employment.

Fig. Industry of Textile

The textile and apparel industry production growth is still at 2 per cent. Saving jobs is not a useful argument to keep the textile and apparel industry going. Textile and apparel industry segments seen to have completely different structure and performance profiles. For some industry segments, prospects are good, for others, prospects are bad. Employment, competitiveness, productivity and profitability are the critical variables. Prospects are good. for weaving industries, non-woven industries, and the tufting industry. Prospects are common for the throwing, winding, and thread industries and the knitted outerwear industry. Lastly, prospects are bad for the circular and warp knit industries, the hosiery and knit underwear industry, and the apparel industry, in general. The following are the reasons for good prospects of certain industries. The weaving industries invest heavily in new technology which increases productivity and profitability.

The technology is chiefly foreign development. The non-wovens industry is clearly in the early stages of the industry life cycle with heavy exports and increasing employment, whereas the tufting industry is fragmenting due to negative returns to scale. This implies that larger firms do not enjoy higher productivities due to their large size. This has, however, not led to decreases in employment. Exports are extensive.

Industries with average prospects are the knitted outerwear, throwing, winding, and thread industries. The knitted outerwear industry has high positive returns to scale in the productivity sense. This reasons consolidation, but with no effect on employment. The throwing. winding and thread industries substitute lower cost capital for high cost labour with increasing automation. There will be further investments in new technology in chiefly large scale plants. The industry is competitive. One can expect decreasing employment due to technologies.

Mainly bad prospects face the apparel and knitting-related industries. The hosiery and knitted outerwear industries are disinvesting due to imports and not technology. In the circular and warp knit industries, both technology and imports reduce employment.

For instance, highly flexible and innovative marketing of sweaters by foreign firms has gained a large share of the sweater market, but the foreign firms have no technological advantage over firms. Besides, technological advances do not improve the financial performance. The apparel industries are depressed by large imports. New technology reduces the labour content, but does not always improve productivity. In general, the textile industries can benefit from investments in new technology which is chiefly produced in Europe and Japan to improve productivity and profitability with a reduction of labour content.

The apparel industries can reduce labour content to a certain point. However, this has not, so far, importantly improved productivity. In the aggregate, both industries will face large decreases in labour content. Some apparel industries must advantage from their potential in marketing by getting international marketing expertise. There are potential growth areas in the textile and apparel industry. Productivity in the textile industry has increased faster than the industry average over the past twenty years. The potential for further increases in productivity is high, but increased productivity does not essentially yield short run financial improvements. Long-term planning is required, and a particular independence from the financial markets is beneficial. Privatization of textile companies is very common. Productivity increases are mainly geared towards reducing the labour content in production. Higher productivity machinery is chiefly supplied by overseas manufacturers. The lost its competitiveness in the development of textile machinery.

Further developments include system-wide improvements in materials handling and more efficient work flows rather than just invention of new gadgets. Productivity increases in the apparel industry are slow. They all emphasize the reduction of labour content. The sewing process is the limitation lo higher productivity increases.

There have been no revolutionary improvements in systems to join fabrics, including sewing systems. System wide improvements promise higher returns than "futuristic" technologies. Improved management and control systems can drastically improve productivity of the textile and apparel industry, overall. Inventory control, the optimization and speeding up of order-to-delivery time are deasive. The time from order placement to delivery can be decreased by improved cooperation of tiw difficult industry segments. The fragmentation of the industry must be decreased in the sense that firms integrate vertically to achieve economies of scale in productivity and flexibility with fast response to the market. Vertical integration is not necessary in the sense of finance and ownership. However, this means that hostility among the industry segments disappears.

Most system-wide improvements require coordination between the jndustry parts without which the full impact of these technologies will not be realised. Productivity in the textile industry is the most dominant factor in the competition among producers. The entire productivity is somewhat less important in the apparel industry against foreign competitors since there are other factors like exchange rates and high labour costs which dominate. Productivity advantages over competitors can only be reached with regular upgrading of machinery. Since the technologies in these industries, particularly in the weaving and spinning sectors, change rapidly, the machinery becomes obsolete in short time periods. The capital needs, for example, for approximately five year cycles in the spinning industry, are enormous. Only financially strong industries are able to upgrade their production technologies without assistance.

There are three important technology related factors which keep the textile and apparel industries from being competitive and dominant in the international markets.

- The textile and apparel industries receive the same technologies as do competitors in the world markets. The textile and apparel industries have no technological benefit over their foreign competitors.

- The textile and apparel industries are dependent on foreign suppliers for machinery.
- The textile machinery industry is fastly declining and has lost ground in chief machinery markets in some areas, such as spinning and weaving. There are no substantial American machinery producers in the market in these two areas.

The development of textile technology cannotes that productivity in the textile industry is above the average of all manufacturing industries. Unfortunately, this is not true for earnings. The technologies, however, are not to blame for the lack of performance. For instance, loom speeds have almost doubled since 1970.

Tighter international markets, imports, and the Tack of appropriate marketing are more likely the cause of unsatisfactory financial performance of the textile and apparel industry. The use of the newest technologies is not restricted to industrialized nations.

Developing nations are rapidly increasing their capacities with the newest technology available. Machinery manufacturers are likely to shift their emphasis to the Far East, due to higher growth prospects. Small advantages in productivity which enable firms to be more competitive are the key to an industry which competes on a cost basis.

Labour cost differentials still are a high disadvantage for manufacturers. However, a positive aspect is that in several segments of the textile industry capital intensity is very high and labour costs become less important due to automation. New technologies in the textile industry reduce labour content, and improve productivity. In the apparel industry, the labour production component is reduced without substantial productivity increases.

The limits to labour reductions in the apparel industry are set by the sewing operation. It is feasible to almost fully automate some textile processes. One can hope that the textile industry will be more successful in becoming capital intensive than the apparel industry. The technology is available, and there are considerable developments going on in all parts of the world. Generally, apparel technology still lags substantially in automation.

The development of technologies is not at an end. Automation and robotics will further increase productivity and reduce labour. During the past decade, however, the industrialized nations, lost ground in the world markets despite increases in productivity.

While capacities were only unused in developing nations up to now, they increasingly attain full capacity utilization. There is a high potential for increases in the production quantities of developing countries. Major issues on behalf of the textile competitive situation are related to the level of the technologies in the different industry segments. The question is-whether textile mills have technologies on the newest level.

In general, the level of textile technology is considerably lower than in Europe, but both continents have superior technology levels to the developing nations. The gap, however, is fastly closing.

Developing nations, especially Korea, Taiwan, and Hong Kong are spending heavily on the newest technologies at a higher rate than the Western nations. These three Far Eastern countries can no longer be counted as developing nations anymore regarding their textile and apparel industries. The weaving industry is the largest spender on new technologies. These investments are concentrated among the hin biggest corporations.

The ten largest companies account for 85 per cent of these expenditures. It appears that the upgrading of machinery is limited to large scale operations. Apart from upgrading textile and apparel machinery, there are three traits An current development that go through all the various manufacturing processes. These are the automation of all processes, monitoring of the complete manufacturing process, and the integration of all manufacturing technologies in flexible units and fewer steps.

Productivity increases and reductions in the labour content are more likely feasible through-further develop-ment of the current technologies and system-wide improvements than through expensive "futuristic" technologies like molding garments to individuals in zero-gravity environments. Most textile machinery is imported EEC countries and Japan. Imports of textile machinery are concentrated in the fabric and yam industries as well as in some other industries.

The American market is still the most important market for all machinery producers. However, some Swiss machinery manufacturers realize the limited growth in the textile industry and are shifting towards developing countries. West Germany and Switzerland together account for two-fifths of the world exports in textile machinery. Czechoslovakia and Japan have emerged as important competitors. The concentration is rather high in the machinery industry.

The Swiss based Sulzer corporation recently acquired its strongest competitor, Ruti. Moreover, Sulzer-Ruti has close ties with Toyoda, the Japanese leader in looms as well as to British air-jet technology. These impressive market positions have been achieved with high R&D in their respective fields and traditional reliability in service.

The American machinery producers basically lack any momentum in this regard. Most of the European success is also because of their service. All the big machinery producers have service facilities in the heart of the textile industry. The whole staff speaks English. If essential, spare parts can be flown in within a reasonable amount of time. The success of the American textile industry depends heavily on these machinery manufacturers. However, this increasing dependency should be a major concern of the American producers.

A key to competitive textile industries is to lower down the diffusion of new technologies. This would mean to emphasize in-house, proprietary technology. It is reasonable to believe that the textile companies are unable to invest in process R&D. Furthermore, the market positions of the textile machinery producers is such that a natural dependency is inevitable.

The apparel industry, however, can take advantage from developments in the electronic industries for further applications of computer controls. American resources are enormous, if used. In general, the textile and apparel industry is largely dependent on foreign machinery which is sold all over the world. The structure and performance of the various industry segments in the textile and apparel industry vary widely. New technologies influence the structure and performance of these industries differently due to the variety of the competitive forces.

Also, the structure of the industries is likely to change rapidly with the introduction of new technologies. Past data indicates that there have been important differences between the textile and apparel industry's profitability. The textile industry's profitability has been significantly below the apparel industry's profitability.

A reason for the differences in earnings might be in the higher cyclicality of the apparel industry's earnings. Apparel purchases are more volatile to the business cycles than the primary textile industry. One might hope, therefore, higher fluctuations in apparel earnings. Earning changes in the apparel industry influence the textile and fiber producer's earnings with a time lag but somewhat smoothed out. On the other hand, there are no differences in the average productivity of labour between the two industries. Contrary to popular belief, the apparel industry is not less productive than the textile industry. This is even more interesting when the different levels of capital and labour intensities are compared.

The textile industries are significantly more capital intensive. As a consequence, productivity is not an argument for the observed reductions in the labour force. It is quite the escape from cost disadvantages in the U.S., thus the labour cost differential. Productivity increases come as a "free gift" with further automation and robotics. Among the textile and apparel industries, onc can observe significant differences in the levels of labour intensity, capital intensity and technological volatility. The apparel and textile industry's levels of labour intensity are different, but not as widely spread as the capital intensity levels.

Gross value of fixed assets per employee is about five times as large in the textile industry as in the apparel industry. The capital requirements are, therefore, much higher in the textile industry. Besides, this enhances the industry's entry barriers. The apparel industry's entry barriers are very low, whereas, in the textile industry, there are considerable capital related entry barriers and scale related effects. Industry concentration in the textile industry is much higher than in the apparel industry. In the textile industry, there are more large firms dominating the market, whereas the apparel industry cooprises, smaller, but so many firms.

Interestingly enough, the industry's firm size adjusted concentration, and the scale assessment levels of the two industry groups are not importantly different. When the firm size differences are ruled out, the concentrations of the four largest firms are not different for the two industries. This implies, that in both industry groups, there are a few dominant firms on the market. The respective concentration levels in both are high.

In both industries, the four largest firms account for 50 per cent of the respective markets. The relative levels of economies of scale in the two industry groups are similar. It is not the case that in one industry chief large or small scale operations dominate. However, one can observe a wide variety of economies of scale in both industries.

Regression analysis indicates that in the apparel industry, monopoly benefits (excessive profits after reinvestments and transfers to the shareholders) are literally not responsive to changes in the labour intensity. In the textile industry, monopoly profits are negatively related to labour intensity. These results clearly indicate, that for the apparel industry, the labour intensity is not directly liable for the profitability performance.

Apparel industries that use more labour as their production factor are no better off than apparel industries that are relatively more capital intensive. On the other hand, in the textile industry, the more labour intensive production facilities are clearly worse off. As a result, other factors, for example, the increasing competition from imports, are more likely to influence the profitability of the apparel industry. Obviously, apparel industries of various degrees of labour intensity have been equally capable of protecting themselves from competition.

Textile industries which are less labour intensive benefit from their technological benefit. This is reflected in the fact that they are more likely to behave in an oligopolistic manner, and thus, with higher concentration.

The trend in the textile industry should, therefore, be in increased expenditures in capital to constantly upgrade the technologies. Generally, industry concentration is invariant to profitability. Firms with big market shares face the same struggles in generating enough monopoly advantages as small share firms. The tendency, however, is that the apparel industries' profitability more positively responds to industry concentration.

Thus, apparel firms with high market shares tend to be more profitable and are more successful in the battle against imports than apparel firms with small market shares. The amount of economies or diseconomies of scale has no effect on the profitability of the industry. This means that the behaviour of the industries with respect to enlarging or downsizing the operations has no effect on the profitability.

The argument that larger plants are more competitive than smaller ones does not hold in both industries with respect to profitability. It is explicit from the industry structure profiles that no textile or apparel industry is identical in the way it can and will exploit new technologies, especially automation and robotics. Not all textile and apparel industries are able to drastically improve their competitive position with new technologies. In these cases, market forces are such that imports are at levels where essentially small improvements in cost and product differentiation have only marginal impact.

This is where critical market shares have already been captured by foreign firms. The most apparent examples are in knitting and in the whole apparel sector. In the relatively capital intensive industries, like the weaving and spinning industries, automation can and will enhance the competitive position of firms which can afford it. Also, the necessary and feasible technological changes are closer at hand in the latter industries, to be beneficial within a reasonable amount of time.

Winding technologies took big steps in the past years. About full automation is or can be reached soon. The spinning process still lags in full automation. However, immense improvements are under way in the basic spinning technologies. Rotor and air-jet technologies eliminated steps in the production process. The reduction of steps in the textile production process is the most important source of cost reductions. New weaving systems drastically speed up the fabric forming process.

High segmentation of the textile machinery industry, however, hinders; efficient development of joining technologies in the sense of improving the overall textile manufacturing process. Technological developments are merely further refinement of current machinery. Completely different ways are pursued by the newly emerging textile industries, the tufting and the non-woven industries. As in every early stage of the industry life cycle, proprietary technologies are developed in-house.

The profits are apparent in the impressive domestic and international competitive positions of the U.S. firms which are at least partly a result of efficient technologies. The apparel industries have yet to overcome the technological disadvantage of the needle-sewing process which needs as much labour as sewing stations. Rumors of prospective revolutionary sewing robots tend to neglect the commercial value of such gadgets. Unless these technologies are cheaper than labour, their practical value is not evident.

Therefore, materials handling is where efficient technological improvements have economic value. New technologies are likely to change the economies-of scale in manufacturing of the respective industries. The weaving industry is the most clear example. Air-jet weaving technology drastically enhanced economies of scale. However, if this happens in an uncertain market environment where smaller but various production batches are postulated, flexibility might be advantageous.

The apparent tradeoff between flexibility and big economies of scale is the subject of substantial research in suitable weaving technologies in Europe and in Japan. Changes in the industry structure can be hoped, once the development of joining technologies is at the Point where spinning and winding processes, as an example, can be performed without intermediate steps. One might hope more vertical integration caused by more efficient materials flow between the spinning and the winding process.

Other instances are likely to have the same effect as weaving directly from spinning frames. These ideas obviously are yet to come, but they will change the industry structure. Specialists will be less in demand and firms will vertically integrate to coordinale the workflow It is also expected that more vertical integration raises entry barriers. If this is not too late foreign competitors might find it more difficult to enter the markets. On the other hand larger economics of scale distract from essential flexibility in the marketing and manufacturing sense.

It is potent not to be undermined by quicker foreign suppliers. The degree to which automation is introduced must be carefully evaluated with respect to changing economies of scale flexibility, entry barriers and overall industry attractiveness. The mature textile and apparel industries operate with technologies which are chiefly foreign produced. Weaving and spinning machinery industries are concentrated in Europe and Japan. Diffusion of technologies can hardly be halted, such that first mover effects are crucial for the competition, but their effects are not long lasting.

These technologies can not be characterized as mature.

For instance, the introduction of the Sulzer projectile loom resulted in changes of industry structure. On the other hand, apparel technologies are mature. Technological development is slow and improvements merely concentrate on current technologies. Automation and robotics benefit the various textile and apparel industries to different degrees.

Mature industries merely require heavy automation and robotics to gain cost differentiation. In the growth industries gains in product differentiation and cost benefits can be reached with process innovation and the development of completely new technologies. Changes in technologies are most likely to affect industry structure. The most apparent tradeoff must be made between necessary flexibility and large economies of scale. There is a compelling need to keep closer to changing markets. The volatility of the markets is likely to increase.

Demographics and higher income. lead to higher demand, but the higher demand is more likely in higher quality and variety. This will result in reduction of the demand for "commodity like" products. Proactive product development through understanding of the Gonsumer's needs will dominate. This is where the producers appear to lag behind several foreign suppliers. Some of the foreign suppliers seem to understand the market better than the American producers.

Certain industry segments prove that there are potential growth and export spheres in the textile and apparel industries. Economies of scale in marketing, for instance, through the cooperation of the fiber industry's international marketing expertise and the consumer product or industrial product firms, need to be established. An industry capable of holding international markets may look completely different from today's in its interest in research and the ability to absorb technical ideas, the sophistication of its management in production and marketing, and the way it uses its labour.

Although it is impossible to predict for the entire industry what the future holds in store, it is feasible to predict what will happen to certain industry segments under different environmental conditions. If the textile and apparel industries receive long-term protection from foreign competition where there is little motivation for change, it seems that these industries will continue to invest in new technologies in order to decrease labour inputs, but there is no proof that increased research and development will be undertaken to produce the new technologies.

Further, there will be no motivation to change the "arms-length" or adversary relationships which exist between key elements in the channel such as textile mills and apparel plants. The history of all industrialized nations shows that long-term, open-ended, blanket-type protection from competition places the undue burden of very high prices on consumers and thus more problems than it solves for the nation. On the other hand, removal of all barriers to the free and open entry of textile and apparel goods to the market, while continuing restrictions like minimum wages, safety regulations, and anti-trust regulations on the producers will not serve the national interests as it would produce a fast loss of over one-half million jobs and plunge the nation into a serious recession.

Further, such action could result in the inability of the domestic industry to meet the requirements of the military services in time of national emergency, as has already happened in the footwear industry.

CHARACTERISTICS OF THE MANUAL

In order to promote energy conservation in the textile industry, this manual analyses in detail the actual conditions of energy consumption in each field. Thus, it has been compiled to serve as reference material for the practical application of these techniques which are always kept available for the engineers. A step-by-step description is included as to how to implement the energy conservation measures, specifically in the area of the dyeing and finishing process, where many small- to medium-sized companies are operating.

We strongly hope that this manual will be used as a guide to promote energy conservation in the textile industry and to rationalize the management.

CHARACTERISTICS OF ENERGY CONSUMPTION

TYPES OF ENERGY USED IN THE TEXTILE INDUSTRY

In general, energy in the textile industry is mostly used in the forms of: electricity, as a common power source for machinery, cooling and temperature control systems, lighting, office equipment, etc.; oil as a fuel for boilers which generate steam; liquified petroleum gas; coal; and city gas.

Table compares the energy consumption shares of various specialized technical fields and it can be seen that energy consumption is relatively high in the fields of dyeing and finishing, fibre production, spinning, weaving and clothing manufacturing.

Table. Energy Consumption Share of Each Specialized Technical Field in the Japanese Textile Industry

Unit: Million Yen

Specialized Technical Field	Fuel	Electricity	Total	Share
Fibre production	32,551	21,498	54,049	21.0%
Spinning	3,224	44,262	47,480	18.4
Twisting	291	1,660	1,879	0.7
Textured-yarn production	120	1,543	1,663	0.6
Weaving	4,467	24,848	29,315	11.4
Knitting	4,059	11.709	15.858	6.1
Dyeing	37,661	28,412	66,073	25.0
Clothing manufacturing	8,240	15,420	23,660	9.2
Others	5,959	12,000	17,959	7.0
Total	**96,500**	**161,442**	**257,942**	**100.0**

Note: Calculated from the Tabulated Industrial Statistics

Table summarizes recent trends in the use of various energy sources in the fibre production and dyeing and finishing divisions of the textile industry, where the energy consumption ratio is relatively high.

Technical Field	Fiscal Year	Electricity (10^3 KWh)	Kerosene (KI)	Gas Oil (Ki)	Fuel Oil A (KI)	Fuel Oil B (KI)	Liquefied Petroleum Gas (t)	Coal (t)	City Gas (10^3 m^3)	Total (10^6 Kcal)
Fiber Production	1986	3,839,760	4,067	10	6,182	16	719,547	712	517,177	15,631,924
	1987	889,582	5,587	17	2,221	..	635,941	699	706,402	15,854,930
	1988	3,882,074	3,821	14	12,209	..	622,291	727	788,698	16,012,279
	1989	3,887,386	2,736	3	28,520	..	544,848	656	838,510	15,458,822
	1990	4,066,002	3,571	5	35,503	..	549,098	662	871,265	15,358,554
Dyeing	1986	1,325,692	24,250	400	39,770	11,751	816,942	83,678	7,187	12,103,452
	1987	1,351,241	22,565	406	50,351	11,906	806,477	87,978	30,084	12,322,468
	1988	1,343,959	21,879	442	97,509	12,323	775,313	95,349	37,846	12,610,598
	1989	1,397,260	21,470	471	180,880	10,507	708,562	104,125	37,580	12,706,610
	1990	1,398,127	20,702	614	199,846	10,068	678,942	109,281	32,827	12,724,916

Note: Calculated from the Annual Report on Textile Statistics. Dyeing Represents the dyeing process including finishing, for both woven and knitted materials.

PROMOTION OF ENERGY CONSERVATION TECHNOLOGIES

While the significance of energy conservation awareness is relatively easily understood at home, when a programme is introduced into a factory to promote it, its thorough implementation tends to be delayed at an early stage. Therefore, for its actual course of implementation, it is desired to devise company-wide coordinated measures similar to QC activities at factories. Also, in order to promote energy saving measures efficiently, it is found to be effective to separately consider general management techniques for "rational use of energy" and process-specific techniques to be developed in each specialized technical field.

ENERGY CONSERVATION MANAGEMENT TECHNOLOGIES

Organizational Rationalization

Since energy management is relevant to a wide range of departments within a company, it is necessary to enhance the awareness, improve the knowledge and obtain the participation and cooperation of everybody involved in the production process. Therefore, while it is necessary for engineers and technicians with specialized technical knowledge to play a central role in energy conservation efforts, the implementation of an energy conservation programme itself should not be left to a handful of specialists or specialized sections. Rather, it is desirable to address the task company-wide, for example by setting up an 'Energy Management Committee'.

Improving Efficiency of Electricity Use

Lighting

Due to its nature of operations, the share of lighting in electricity use is relatively high. After the switch from tungsten bulbs to fluorescent lamps achieved considerable electricity savings, electricity-saving fluorescent lamps have been developed and marketed for further improvements, including those capable of reducing electricity use by several per cent for the same level of illumination. In general, the effectiveness of illumination is influenced by various factors, such as the intensity of light source, the reflection coefficient and shape of the reflective fitting, the layout of the room to be illuminated, interior finish, colour and the distance from the light source. Therefore, it is important to re-examine whether the light source is utilized in the most efficient way and take electricity saving measures, if necessary, such as reducing the number of lamps in use by switching from global lighting to local lighting as much as possible.

Electric Motor

The textile industry uses a vast number of relatively small electric motors. Notably, while a conventional machine was driven by a single motor with the generated mechanical power transmitted to various parts of the machine in a collective manner, many modern machines utilize multiple motors with a control board controlling the movement of each motor, which is directly coupled to a machine part to drive it independently from others.

This is also a rationalized feature in terms of energy saving. However, regarding the selection of each motor, emphasis has been placed on mechanical performance, resulting in a motor with an excessive capacity. This leaves considerable room for reexamination from a energy conservation point of view.

Electric Heating

In the textile industry, electric heating has largely been replaced by other methods for some time in order to achieve cost reductions. However, since electric heating only requires a small initial investment as a result of convenience and simplicity in equipment construction, it is still used for small capacity local heating purposes. Therefore, it is desirable to conduct a comparative investigation into alternative heating methods, such as far-infrared radiation heating, high frequency dielectric heating and microwave heating.

Improvements in Efficient Fuel use

Selection of Fuel

As is described before, fuels utilized in the textile industry have already gone through a switch-over from coal to oil. More recently, efficient energy use is under investigation, including the revival of coal on the way to a further move from oil to liquefied and city gases, while reflecting various fuel prices. In selecting fuels, those with good flue gas characteristics in addition to high calorific value and ease of combustion are desired, so that air pollution can be prevented as much as possible.

Selection of Boiler

By and large, boilers used in the Japanese textile industry have experienced a change from Lancastrian- or Scotch-type tubular or smoke tube to water-tube boilers. As a result, boiler efficiency has improved from the conventional 60's to 70's of percentage points to as high as the 90's. Since high performance boilers are prone to a rapid growth of scales inside their water tubes, feed water management becomes important. Furthermore, these boilers have small amounts of retained water and high evaporation speeds so that many aspects of their operation are automated, including feed water and combustion management.

Improvement in Efficient use of Steam

Piping

The noted feature of steam use in the textile industry is that the amount of steam involved is not so large but the locations where steam is required are widespread so that steam losses due to heat radiation from steam transportation pipes and pressure drops are considerable. Therefore, for steam transportation over long distances, high pressure and smalldiameter rather than low pressure and large-diameter piping is desired, with pressure reducing valves placed as necessary to regulate the steam pressure at the point of use, thereby curbing heat losses. Also, as pressure losses around bends are great, it is desirable to make their radii large. In order to prevent steam leaks from joints due to the thermal expansion of the pipe, expansion joints should be placed where required. Furthermore, in order to maintain the temperature inside the valve, tank and treatment tank as well as the piping, it is necessary to install them heat-insulated, using appropriate heat insulating materials, so as to efficiently use steam while preventing heat losses.

Steam Accumulators

Since live steam is often used in dyeing factories, fluctuations in steam use during working hours are large. On the other hand, since high performance water tube boilers and once-through boilers are designed such that water retained inside the boiler is very little, the boiler cannot react to momentary and sudden load changes, while responding to automatically controlled slow load changes is not a problem.

In such a case, a steam accumulator can be installed midway through the heat transporting pipe, between the boiler and the heat consuming load, in order to store excess steam when the load is light by transforming it to heated water. This then transforms the heated water back to steam when the load is heavy in order to reinforce supply to the load. This allows the boiler to continuously operate with the average load and is quite advantageous in view of energy saving.

Recycling of Drain

So far, after its heat energy is consumed, steam has been drained off. However, in view of energy saving, it is necessary to collect and recycle the heat energy carried by the drain water.

Utilization of Heat Exchanger

In each production process of the textile industry, the heating and cooling of gases and liquids as media of heat are frequently required. This is done through heat exchange between different fluids, and in order to avoid contamination or chemical reaction due to their direct contact, heat exchangers are used to carry out indirect heating and cooling. It is important to use the right heat exchanger for the intended purpose.

Measuring Instruments and Automatic Control

Energy saving is an operation to grasp the actual situation of energy use in a factory precisely and quantitatively and to carry out improvement measures in order to rationalize and economize on it. While measuring instruments are needed to obtain quantitative data, it will become more and more important to investigate the use of sophisticated measuring instruments based on recent developments in mechanical and electronic engineering, combined with automatic control systems.

ENERGY USE AND RATIONAL USE OF ENERGY IN PROCESS-SPECIFIC TECHNOLOGIES

Progress in production rationalization is achieved through the implementation of a comprehensive set of measures, including energy conservation technologies as the centerpiece measure, along with time management, labour saving, natural resources saving and space saving. It has been frequently pointed out that, along with management techniques described earlier, the improvement and development of process-specific techniques on energy conservation greatly contribute to the rationalization of production. Here, process-specific techniques relating to energy saving are summarized for each specialized technical field.

Fibre Production

Exhibiting relatively large-scale structural forms in the textile industry, this division has already reached a high level of production rationalization; as is well known, it is technologically aiming at diversification into such high valueadded goods as super extra-fine fibre and inorganic functional fibre, commonly referred to as *shingosen.*

In particular, the following techniques relate to energy saving:

- *Raw Material Production Process*: Implementation of energy saving through improvements in the process and reaction conditions
- *Polymerization Process*: Reduction' in polymerization time by means of high efficiency catalysts, polymerization methods, etc.
- *Spinning Process*: Promotion of energy saving through combining the POY (Pre-oriented Yarn: Yarn with some stability with its molecules partially having gone through orientation) and DTY (Draw Textured Yam: false twisted yarn produced while drawing POY yarn) methods and an expanded use in multi-folded spinning yarn.
- *Newly Built Factories*: The factories built during the high growth period have large margins and allowances for production increase so that

high losses would result if production decreased. Therefore, suitably sized factories should be constructed.

Spinning

Regarding technological trends in spinning, moves towards high speed and large package size have been investigated in order to achieve labour saving through as much automation as possible. As a result, energy consumption has been gradually increasing. However, in view of price competition with overseas companies, further labour saving as well as energy saving is desired. Table compares a modern and a traditional factory in terms of electricity consumption for each plant/operation.

Table. Electricity Consumption Per 1000 Spindles for Each Plant/Process

Factory		Modern Factory		Traditional Factory	
Electricity Consumption	Spindles	KWh/1000	% Spindles	KWh/1000	%
Spin-ning Plant	Mixing/Opening	16.7	7.5	4.2	6.0
	Carding	17.7	8.0	4.9	7.0
	Combing	10.9	5.0	2.2	3.0
	Drawing/Roving	9.1	4.0	2.8	4.0
	Ring Spinning	66.1	30.0	34.5	48.0
	Finishing	14.7	6.5	4.7	6.0
	Subtotal	135.2	61.0	53.3	74.0
Air-conditioning	plant	85.9	39.0	18.7	26.0
	Total	221.1	100	71.9	100

Namely, it can be seen that a modern factory as a means of achieving production rationalization requires approximately three time as much electricity as a traditional one, with electricity consumption particularly increasing in the air-conditioning plant. In terms of processing operations, fine spinning, as the main operation of the spinning process, consumes a large amount of electricity. Thus, energy saving measures are required in these fields.

Ring Spinning Operation

For the fine spinning operation, electricity is consumed in driving the spindles, packaging, spinning, drafting, and operating the lifting and cleaning mechanisms. It is desired to curb the increase of electricity consumption as much as possible by setting an optimal condition for each of these electricity usages.

Air-conditioning

Although as an ideal working environment a room temperature less than 30°C is desirable, in cases where the working environment has been drastically improved in most other aspects with work load also reduced, a slightly increased room temperature may be permitted. As has been reported, there was a case where raising the regulated temperature from 30°C to 32°C resulted in a reduction in the electric power demand of a carrier with a contract demand of some 8,000 kW by 190 kW.

Also, there are many instances of seasonal switch-over from a damper to a pulley as a means of readjusting the blown air volume; this is in order to recycle the air sucked from the processing machine for each operation through a filter back to the same room, and it is therefore necessary to recheck the locations of fans for suction and returning.

Textured-yarn Production

While synthetic-fibre textured-yam is mostly produced with false twisting machines, its history of rationalization is characterized by challenges for high speed operation. As their operating speeds increased, driving and heat-curing motors and and other peripheral equipment became larger, accompanied by an inevitable increase in electricity consumption. Although this may be acceptable as long as the production improvement resulting from a high speed operation covers the increase in electricity costs, reductions in energy cost would surface as an avoidable urgent task, should a sharp increase in electricity charge occur.

It can reasonably be said that the major form of energy consumed in the production of synthetic finished-yarn is electricity. Although the amount of electricity consumed in each piece of equipment varies with factory scale and the type of false twist machine, and therefore cannot be treated in a standardized manner, generally accepted average values may be taken as 3.5 kWh/kg for a single heater system and 5.0 kWh/kg for a double heater systemas one report suggests.

Of all the energy consumed in finished-yam production, 70% is accounted for by false twist machines. Table shows a breakdown of this energy consumption:

Table. Example of False Twisting Machine and Electricity Consumption (kWh)

Processing Machine spindles)	**Single Heater (216 spindles)**		**Double Heater**	**(192**
Equipment	**Capacity**	**Utilized**	**Capacity**	**Utilized**
Main motor	15.0	8.0	13.5	9.5
Exhausgasmotor	1.5	1.0	2.2	1.5
Yarn sucking motor	2.2	1.5	–	–
No. 1 heater	32.0	16.0	15.0	7.5
No. 2 heater	–	–	11.0	6.0
Total 50.7	**26.5**	**46.7**	**24.5**	

Sixty per cent of all energy consumption by a false twist machine occurs in the heater. Therefore, improvements in the heat insulation of the heater and the lowering of heater temperature may be considered as energy saving measures. Since the latter has implications in the characteristics of the finished-yam, whether or not it is adopted should be examined on such occasions as in the development of a new product. Since air-conditioning plants are designed based on the conditions applicable at the time of installation, it is desirable that they be re-examined against the present conditions.

Weaving

Rationalization in fabric production is such that while various improvements in machinery aimed at high speed operation and labour saving have been carried out, the amount of energy use per unit of the product has gradually increased. Regarding loom design, high productivity shuttleless looms such as water jet, rapier and gripper types have successfully been introduced, with air jet models put in practice in the production area of industrial fabric material. The amount of energy consumed by each loom during its weaving operation can be estimated from the motor capacity and weaving speed.

Conventional shuttle looms are based on the weft-insertion method, incorporating a shuttle zooming to and fro with a large inertia mass and mounted with extra weft, and they also use energy consuming pirns as an integral part of the machine. For this reason, the shuttleless looms' contribution to energy saving cannot be regarded as too high. On the other hand, as a large amount of energy is consumed in sizing, as one of the preparatory operations for weaving, the introduction of foam and solvent sizing operations are being investigated. Furthermore, long fibre fabrics using nonsizing filaments have been developed, eliminating the sizing process altogether.

In a reported example, the introduction of a new heat exchanger into a sizing machine with a very poor sealing capability achieved more than 40% of energy saving.

Knitting

The share of energy cost in the total cost of production is not necessarily high for the knitting process. However, of the main production facilities for this process, knitting machines have also been undergoing a shift towards high speed and large capacity and fine gauge features; the current industry trend is for high added-value goods and multi-line, small-volume production based on advanced systems such as computer-controlled pattern making mechanisms. Therefore, a potential tendency for increased energy consumption should be taken into account.

As a result, it is desirable to conduct a comprehensive re-examination of the production schedule along with the implementation of actual energy conservation measures in order to reduce or restrain the share of energy cost in the total production cost.

Dyeing and Finishing

It is very important to advance energy conservation in the dyeing and finishing field, which has a high energy consumption share in terms of both the amounts of money and energy used.

The dyeing and finishing process consists of many interwoven unit operations, and it is well known that the process generally goes through repeated wet and dry operations.

The heat balance of a unit operation can mainly be considered as the difference between the total supplied heat on the one hand and the sum of the heat required by the system and various forms of heat losses on the other. Figure graphically summarizes the major factors in a thorough implementation of energy savings.

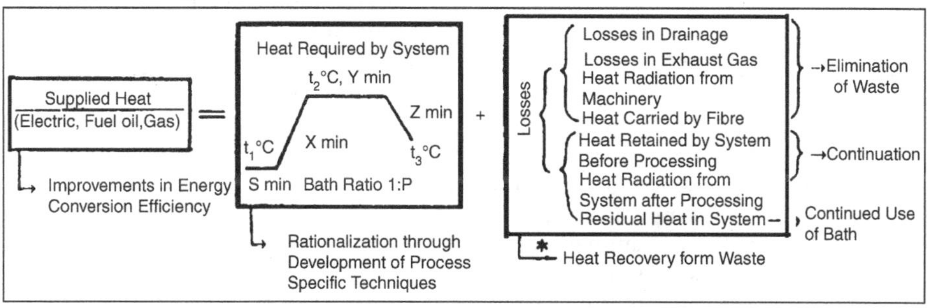

Fig. Heat Balance in Unit Operation

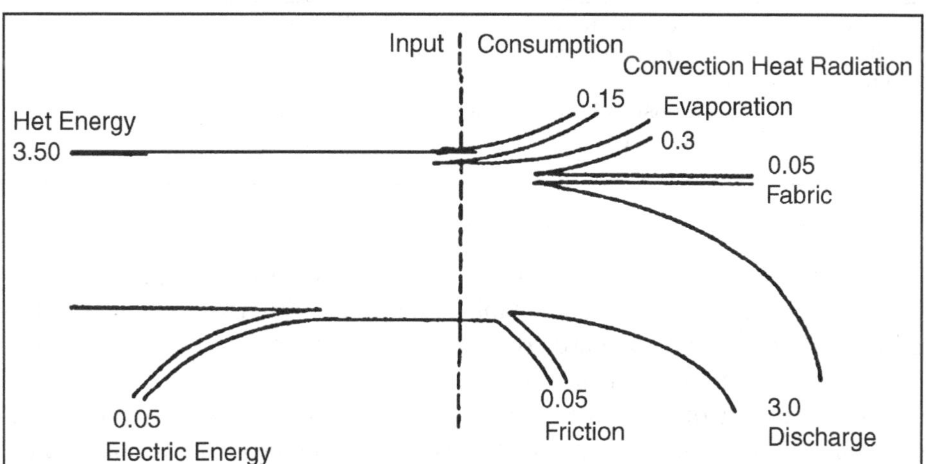

Fig. Shows an Example of Heat Balance in a Continuous Water Cleansing Machine.

This clearly illustrates the importance of the development and utilization of processspecific techniques, apart from the already-described management technologies. Table shows that the implementation of production rationalization eventually relates to energy conservation. The following are brief explanations of typical examples.

Table. Relationship between Production Rationalization Techniques and Energy Savings

Produciton Rationalization Technique	Mechanism	Effect
Time saving	(1) High speed processing of unit operations (2) Reduction in waiting through an (3) Elimination or merger of unit operations	Reductions in energy use Per unit operations time between unit operations improvement in productivity
Labour saving	(1) Implementation of automation (2) Strengthening colorimetric management	Reductions in the frequency of reprocessing through a reduction in the failure rate
Energy saving	(1) Reduction in bath ratio (2) Reduction in treatment time (3) Reduction in margin of temperature rise (4) Re-examination of drying method (5) Switch to non-water-based operations	Reductions in energy cost
Conservation of natural resources	(1) Utilization of continuous bath	Utilization of system's residual heat
Space saving	(1) Construction of modern factories	Improvements in factory-wide energy saving effects

Elimination or Merger of Unit Operations

The currently employed dyeing techniques are based on unit operations which have been developed and established for use with natural fibre. For this reason, the traditional standard treatment steps are often applied to blended yam fabrics as a matter of principle.

However, through omitting or merging some of the unit operations according to the usage of the product and considering the characteristics of the coexisting synthetic fibres, it becomes possible to achieve energy conservation.

Table. Unit Operations in Preparatory Process of PE/C Blended Fabric and its Processing Characteristics

Combination and Ordering of Operations	Number of Units	Processing Characteristics
Singeing→ Desizing→ Drying	10	(1) Cumulative Method

→ Heat setting → Scouring → Mercerizing → Drying	(2) Strict Control (3) Excessive drying operations
Singeing→ Desizing→ Scouring 8 → Bleaching→ Drying→ Mercerizing→ Drying→ Heat Setting and bleaching	(1) Fewer drying operations (2) Difficulty in controlling scouring (3) Great number of wrinkles form processing
Heat setting→ Singeing→ 8 Desizing→ Scouring→ effectiveness in Mercerizing→	(1) Difficulty in desizing (2) Insufficient Bleaching→ Drying→ Drying processing (3) Wrinkles from processing prevented
Singeing→ Desizing→ Scouring 8 → Bleaching→ Drying→ Heat setting→ Mercerizing→ Drying	(1) Satisfactory desizing (2) Fewer drying Operations (3) Caution against wrinkles from processing
Singeing→ Mercerizing→ 7 Desizing→ Scouring→ Bleaching→ Drying→ Heat setting	(1) Fewest drying operations (2) Difficulty in desizing (3) Small mercerizing effect

Reduction in Processing Bath Ratio

It is easy to understand that a reduction in water use will contribute to energy conservation in the dyeing process which consists of various wet treatment and drying unit operations. It is especially desirable to curb the water consumption because it is linked to the overall water supply cost including that of drainage.

For the reduction of the processing bath ratio, it is necessary to investigate the following measures:

Treatment with Low Bath Ratio

In general, dyeing and finishing methods are classified into the batch and continuous processing methods, and it is recommended to use the latter method where a low bath ratio is desired.However, depending on the details of processing requirements, there are often instances in which the batch method has to be employed.

In such cases, batch processing machines which allow lower bath ratios such as the jigger, wince, beam, pad roll and jet flow types should be selected as far as the circumstances permit.

Figure shows the relationship between the bath ratio and the production cost for the wince dyeing of a cotton fabric with a reactive dye (bright red, medium shade). It is easy to see that the bath ratio has a direct influence on the production cost

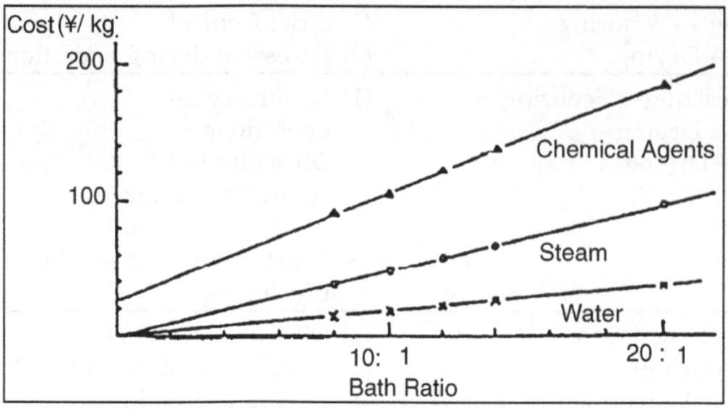

Fig. Relationship between Bath Ratio and Production Cost in Reactive Dyeing

Utilization of Low Add-on Equipment

Several types of processing equipment with a mechanism to uniformly apply the fabric with a minimum amount of liquid necessary in semicontinuous and continuous processing systems are known to be typical examples of energy conservation techniques.

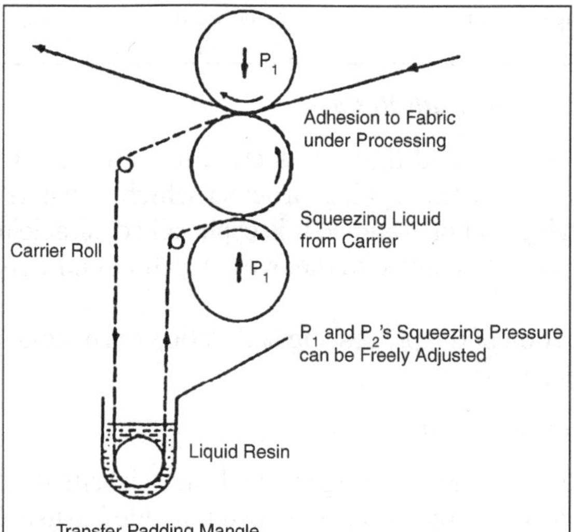

Transfer Padding Mangle

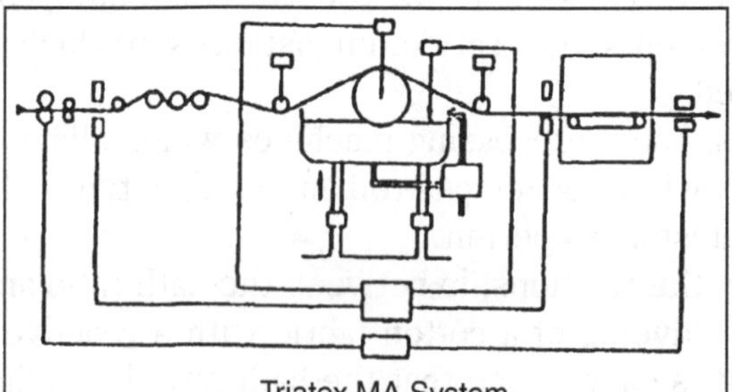

Triatex MA System

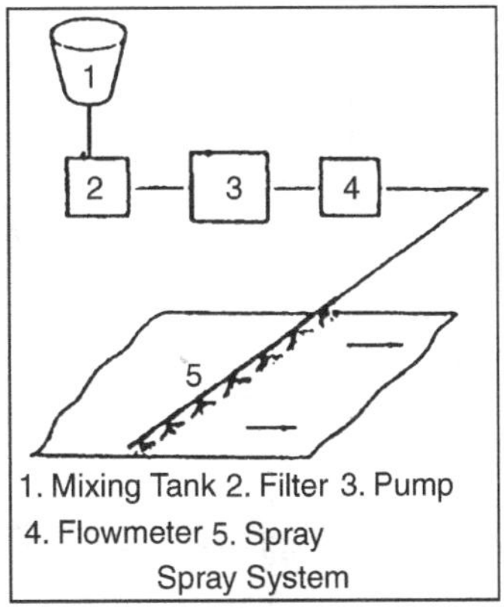

Fig. Typical Low Add-on Machines

Utilization of Low Bath Ratio Processing Equipment

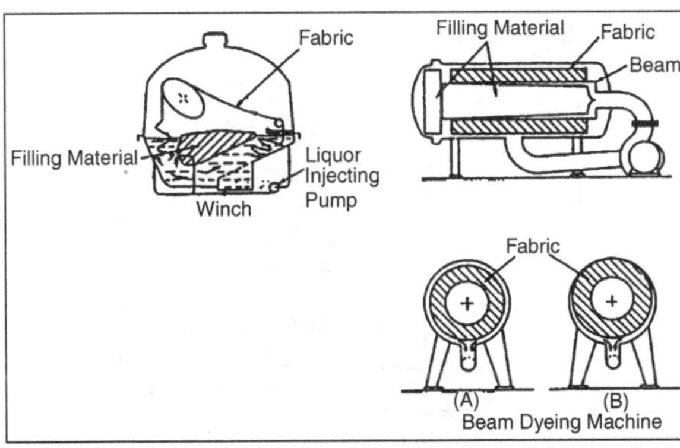

Fig. Example of Low Bath Ratio Operation of Existing Processing Equipment through Insertion of Filling Material

In order to use a lower bath ratio with the existing machinery intact, a method to insert a filling material inside the processing equipment has been proposed. It has been reported that with this method, the bath ratio of a wince could decrease from 25:l to 17:1, and for a beam a reduction was possible from 15:1 to 12.5:1, or even down to as low as less than 10:l where the axis of the beam was made off center with respect to the container body. More recently, low bath ratio processing machines which are built in with the above mechanisms have been developed and put on the market.

Extension of Foam Processing Technique

A typical example of foam processing liquid applying equipment. The foam processing technique is used for the preparatory, dyeing, textile printing and finishing processes, with confirmed effects of promoting energy conservation, but it is desirable to examine details of usage and other practical conditions prior to application.

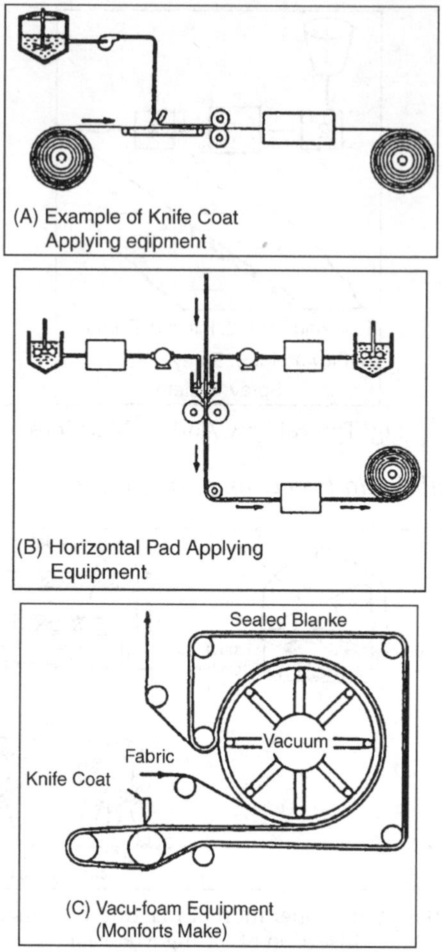

Fig. Typical Foam Applying Equipment

Reduction of Processing Time

As has been pointed out, being a time saving technique aimed at improving productivity, continuous operation with an increase in the size of the processing machine can also further energy conservation. Likewise, for batch processing, the number of technical fields is increasing where the promotion of energy conservation is desired through a reduction in processing time. This tendency becomes more pronounced as the needs of the market become sophisticated. Techniques to accelerate the processing effect with rapid dyeing and plasma treatment are typical examples.

Rapid Dyeing

Rapid dyeing which can drastically reduce the dyeing time and achieve remarkable time savings can also achieve great energy conservation effects when applied to polyester. In order to attain these effects, it is necessary to select dyes with assistants and provide appropriate dyeing equipment. Combined with the foam processing technique, the rapid dyeing technique may also have a potential of leading up to the development of new practical dyeing techniques. Figure shows the position of the dyeing technique in the overall processing bath ratio reduction technique.

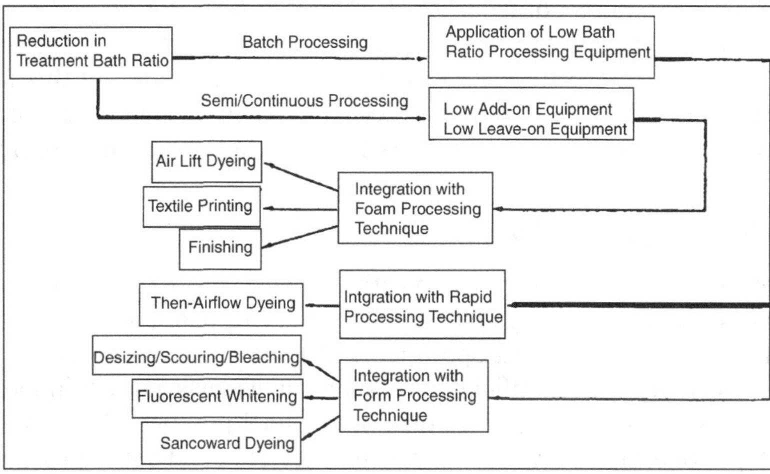

Fig. Situation of Dyeing Technique in Processing Bath Ratio Reduction Technique

Accelerating Techniques for Processing Effects

Aiming at a reduction of processing time, the combined use of a number of new techniques are being studied and it has been reported that processing with plasma, ultrasound, magnetism and radioactive rays accelerates processing effects. Various methods are being investigated to reduce processing time through accelerating processing effects using these techniques in preprocessing, postprocessing, simultaneous processing, etc.

Reduction in Temperature Rise Margin

In many cases, unit operations of the dyeing process are carried out at high temperature. Therefore, to reduce the required margin of temperature rise from heating is very important in view of achieving funda-mental energy conservation along with reductions in processing time. These measures need to be addressed from the following two viewpoints:

Raising Temperature of Inlet Water

If the temperature of the inlet water to be used in the dyeing process becomes relatively higher, the amount of energy to be consumed in raising it to the predetermined value will be reduced. For that purpose, cooperation within a company, or that involving more than one company (using low temperature inlet water at the dyeing factory for cooling purposes, and the high temperature discharge from the cooling system for dyeing purposes) should be investigated, as well as the utilization of natural resources (for example, geothermal and solar energy)

Development and Introduction of Low Temperature Processing Techniques

It is important to continue with technological development aimed specifically at lowering the processing temperature along with raising the inlet water temperature. It would naturally involve the integration of this method with the processing speed acceleration techniques. Low temperature scouring, bleaching, dyeing and curing techniques are some of the practical examples of this.

Re-examination of Drying Method

An important consideration along with the reduction of the processing bath ratio is a re-examination of drying operations. A drying operation is, in principle, inserted after every other unit operation and is an important operation which not only determines drying efficiency as such but also has a direct influence on the morphological stability and texture of the final product. For this reason, various types of drying equipment have been selected and put to practical use, depending on the fibre material and form involved. In view of implementing energy saving measures, it is particularly important to investigate the following three items:

Reduction of Drying Operations in Number

A study of typical preparatory process configurations reveals that drying operations are involved in relatively high numbers, ranging from one to four units or 14% to 40% of all unit operations in the entire process. Therefore, combinations of unit operations should be sought after such that drying operations between standard unit operations can be eliminated as much as possible. It is especially necessary to cut down on drying operations in preparatory processes which will not directly affect the product's performance or appearance quality. However, an operation which would have come after an eliminated drying operation would have to treat wet fabrics, thereby necessitating special measures that would enable wet on wet treatment.

Improvements in Drying Efficiency

It is desirable to investigate possible improvements in the drying efficiency in terms of efficiencies of both the dewatering and drying steps. While it is well known that the most efficient methods of dewatering and drying are by means of a mangle and a cylinder dryer, methods which have been practiced for a long time, they are also known to have limitations in terms of applicable fibre materials and forms.

It is necessary to investigate new drying methods (high frequency drying, microwave heating, far infrared radiation heating, etc.) together with other measures such as utilizing vacuum liquid removal, adding a drying-facilitating organic solvent to the treatment liquid, and combining foam treatment systems with non-foam ones.

Recovery of Heat Energy

Along with active energy saving measures, it is important to carry out the collection and recycling of the energy used in unit operations. The collection of heat energy should start with thoroughly grasping the basic energy balance of each unit operation.

Shift to Solvent Processing

In the dyeing process, although water has been used as the only abundant and cheap resource so far, it is becoming difficult to obtain high quality water in large quantities at a low cost. The worsening of river pollution coinciding with an increase in population density is inevitably creating a situation where the cost of water will gradually increase, including the investment for improvements in waste-water treatment facilities.

In addition, although dry-system processing has been contemplated for a long time due to the fact that most energy is consumed in the heating and evaporating operations, it has to date only been applied to a specific area on a limited scale. However, it is a technique which deserves attention as a promising process in the mid- to long-term future. This technique has the following two variations:

Organic Solvent Processing

While the solvents to be used for dyeing processing are categorized into four main groups—halogenated hydrocarbons, petroleum derivatives, aromatics and oxygen-containing solvents—halogenated hydrocarbons are generally recommended as they do not cause a fire or explosion (provided that thorough countermeasures to groundwater pollution are taken).

It is well known that in terms of energy conservation, these solvents have an advantage over water-based ones in all of these aspects: specific heat, latent heat for evaporation, heat needed for evaporation and evaporation speed. There are a number of proposals for solvent scouring, solvent dyeing and solvent finishing, including those already put to practical use as a differentiating technique.

Inorganic Solvent Processing

Liquid ammonia is one of the agents being considered for dyeing applications as a inorganic solvent. Of its typical processing techniques, liquid ammonia mercerizing and liquid ammonia dying are given particular attention.

Use of Continuous Bath

While textile manufacturing techniques which promote the conservation of natural resources include grease refinement from raw wool, the collection and recycling of warp sizing agents, and the re-use of alkaline waste liquid arising from the mercerizing process in a scouring bath, in terms of energy conservation the use of continuous baths which utilize the residual heat of the system are particularly important in view of energy conservation.

If the continuous use of a processing bath is introduced with the necessary conditions being met, thus allowing only those materials consumed in the dyeing process to be replenished, in particular with the unit operations designed for the batch method, it will greatly contribute to the recycling of heat energy in addition to achieving the conservation of natural resources and the rationalization of countermeasures to water waste.

In the dyeing process with a high heat consumption, the use of the continuous bath deserves particular attention as a technique whose practical application is an urgent task to help implement the remaining rationalization measures.

Space Saving

With an expansion in the practical use of knit mercerizing and ammonia mercerizing, the characteristics of the hot mercerizing technique is also attracting attention. It has been pointed out that, since poor uniformity associated with the traditional mercerizing due to the hydrophobic nature of cotton grey fabric is dissolved in hot mercerizing which uses heated sodium hydroxide, the rationalization of the preparatory process can be greatly advanced. Figure shows an example of comparing some space saving effects achieved by the introduction of hot mercerizing.

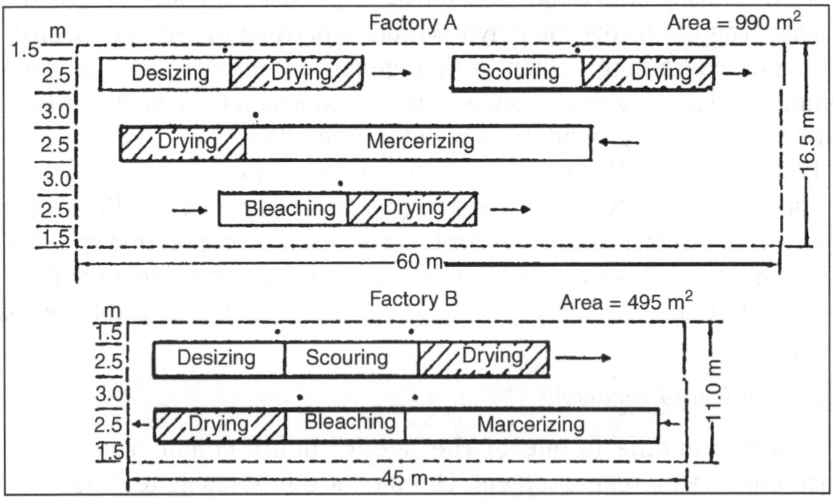

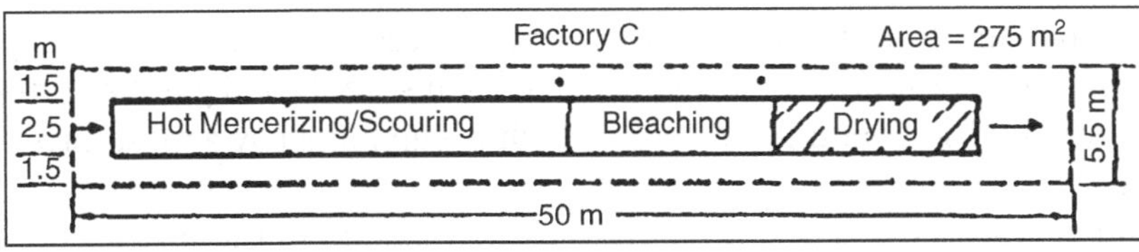

Fig. Space Saving Resulting from Introduction of Hot Mercerizing

Note:

- *Factory A*: Although it has the same types of machines as Factory B, it has introduced a drying operation for each unit operation, emphasizing flexibility.
- *Factory B*: Considers the production rationalization resulting from continuous processing./
- *Factory C*: It has the same production capability as Factory B, but has reduced the preparatory process related to mercerizing.

Table. Compares Operating Conditions of the Same Three Factories

Factory	Investment	Area (m²)	Workers (person)	Electric Power (kW)	Steam (kg/hour)	Water Supply (1/hour)	Waste Water (1/hour)
A	109%	990 (200 %)	14 (155 %)	242 (114 %)	5,727 (130 %)	14,535 (100 %)	14,535 (100 %)
B	100%	495 (100 %)	10 (100 %0	216 (100 %)	4,386 (100 %)	14,535 (100 %)	14,535 (100 %)
C	67%	275 (55.5 %)	5 (50 %)	114 (52 %)	2,500 (57 %)	7,785 (54 %)	4,785 (54 %)

At the time a new space-efficient factory is being built, it is possible to incorporate a programme to introduce such facilities as to be complete with factory-wide energy conservation measures.

Clothing Manufacturing

The energy consumption share of the clothing manufacturing division which consists of large numbers of small-sized companies and their employees in the overall textile industry is not necessarily low but the ratio of energy cost to the total cost is relatively low.

However, the energy cost forecast is inevitably a gradual increase under circumstances where the production of high value-added goods is required, along with the implementation of labour saving measures, as a result of the challenging market environment characterized by personalized and diversified consumer needs, high demand for quality goods, short product cycles, etc. Therefore, it is desirable that a comprehensive rationalization programme be investigated apart from reductions in energy consumption.

STRUCTURES OF TEXTILE MARKETS

In general, it is well known that the developmental process of a textile market moves from a product-oriented stage, where all goods produced are sold to a consumer-oriented stage where the desired goods are only those which will satisfy the consumer's demand, comprising his wants and purchasing power. This is via a mass-consumption stage which is supported by a mass-production system.

The characteristics of textile products demanded in the markets in these different developmental stages may be classified into price-sensitive mass-production type basic clothing, which applies to the earlier stages, and multi-line, small-volume production type fashion clothing which pertains to the last stage and needs to satisfy each individual consumer's taste.

BASIC CLOTHING

Assuming the amount of fabric necessary for people to live a physically comfortable life in a given environment can be derived from the weight of clothes which is enough to keep a certain level of body temperature, given such factors as air temperature, humidity and wind velocity prevalent in the region, basic clothing can be regarded as a cumulative total of such clothes.

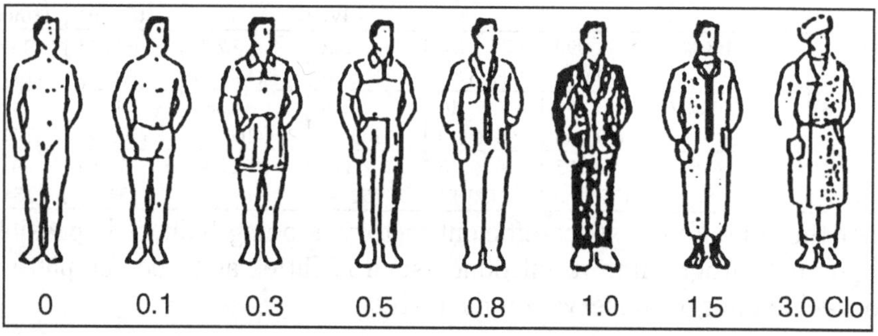

Fig. Typical Relationship between Clothing and Clo Value

Figure Summarizes the Relationship between Some Clo Values and Examples of the Corresponding Clothes. Table. shows the results of calculations to find the weight of clothes required under average climatic conditions in summer and winter in three Japanese cities.

Table. Example of Effect of Climatic Conditions on Clo Value and Weight of Clothing

Region	Season	Temperature (°C)	Humidity (%)	Wind Velocity (m/sec)	Clo Value	Required Weight of Clothing (g)
Sapporo	Jan	–4.9	74	2.3	4.1	4005
	Aug	21.3	79	2.7	1.2	1189
Tokyo	Jan	4.7	53	3.1	3.1	3034
	Aug	26.7	75	3.0	0.6	607
Fukuoka	Jan	5.7	67	3.7	3.0	2937
	Aug	27.3	76	2.9	0.6	607

Notes:

1. The clo values are shown assuming an average skin temperature of 33.5°C.
2. One unit of clo value is 0.155°C/W/m^2.
3. The required weight values are calculated with Hanada's formula for female clothing:

   Y=0.00103W - 0.025

   Where Y: Clo Value

   W: Total Weight of Clothing

Although the ultimately required quantity is difficult to determine, the per capita annual textile consumption (kg/person) can be estimated through multiplying the above weight by coefficient a (obtained from the service life of a textile product and its number of units demanded per annum).

CHARACTERISTICS OF TEXTILE MARKET

Most textile consumption takes place in apparel products, and the general tendency is that where further expansion is intended, it will be carried out through the development of industrial applications in which textile products are used as production materials (ranging from fishing nets, tire cords and canvas cloth to geotextile).

Therefore, in order to study the textile manufacturing industry, it is generally sufficient to consider the production of apparel goods.

WORLD'S TOTAL TEXTILE DEMAND AND PRODUCTION BASE DISTRIBUTION

Figure shows the relationship between world population and the total textile demand.

Assuming a global environment in which world population will grow from the present 5.4 billion to 10 billion in 2050, and further to 11.6 billion in 2150 when it is expected to reach a static state, the total textile consumption is forecast to double, even using the current figure of per capita annual average textile consumption (8kg/person).

The textile industry is traditionally regarded as a typical labour intensive industry developed on the basis of abundant labour supply and has a tendency to expand to overseas markets once the domestic demand is satisfied, as illustrated by examples of the established textile industries of many developed countries.

For this reason, even when the textile industry of a specific country is to be examined, it is widely recognized that a business strategy taking into consideration the global textile industry setup is very important.

Figure shows an example of the schematic representation of such a global setup of textile industries.

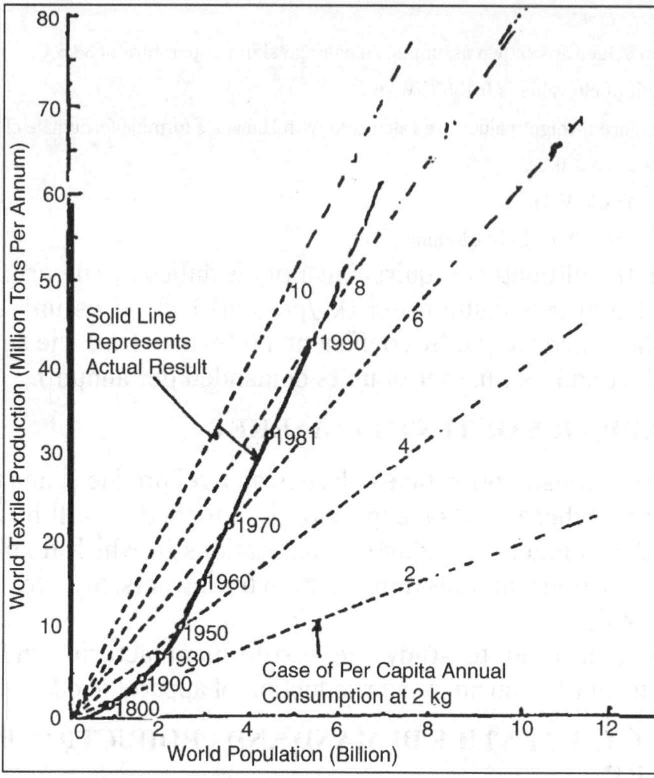

Fig. Relationship between World Population and

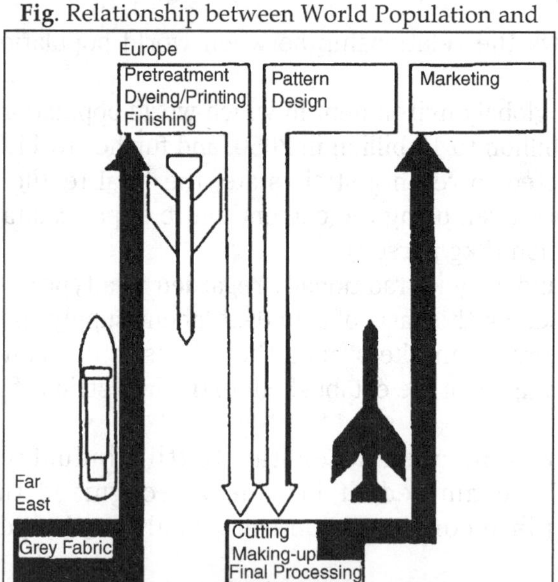

Fig. Example of Schematic Representation of World Textile Industry

Specialized Techniques Necessary for Apparel Goods Production

There are a number of intertwined technical factors involved in various subprocesses or specialized technical fields which make up the overall production process that fabricates and reshapes the raw fibre material into the final textile product to be used by the consumer.

Comparison of Characteristics of Specialized Technical Fields

In Fiscal Year 1989, the textile industry had a share of 4.4% in shipment value. 14.8% in the number of businesses (20%), and 10.4% in the number of employees (23.3%) in the overall Japanese manufacturing industries, characterized by a gradual decline in relative share, although in terms of absolute shipment value it has actually expanded to 1,053% of the Fiscal 1955 level. As has been frequently pointed out throughout its development, the Japanese textile industry has a unique organizational structure consisting of groups of independent companies where all companies in a group belong to one of the specialized technical fields and operate in a horizontal specialization configuration.

A comparison and summary of the major indices of these company groups, classifying them in accordance with their specialized technical fields. From these indices, common company characteristics for each group may emerge. Namely, in terms of business size, the fibre production and spinning subdivisions are in contrast with the rest of the textile industry where relatively large numbers of small businesses coexist. In addition, these groups of small companies can only stay in business by relying on the supply of abundant cheap labour, exhibiting a legacy of the textile industry's past as a labour intensive industry, even on its path towards modernization.

Table compares the energy consumption shares of various specialized technical fields and it can be seen that energy consumption is relatively high in the fields of dyeing and finishing, fibre production, spinning, weaving and clothing manufacturing. As for water consumption, the share of the textile industry in the entire manufacturing industries is 5.2% for fresh water and 1.1% for sea water. Considering the fact that most of this water is used for easy-to-recycle temperature control and cooling purposes, the industry's total water consumption cannot necessarily be regarded as high. However, as is widely accepted, the dyeing and finishing division is placed in a special position in that its water consumption is mainly for processing and washing purposes.

MARKETING TRENDS

This is a marketing research on the Textiles industry and can include information on the background, market structure, definitions, competitors, trends and developments of textiles and is related to other apparel, clothing, yarn, thread and fibre. Global market trends in the textile industry increasingly are being determined by the transition to a quota-free textile trade environment. The quota system that has been in place since the early 1960s ended on December 31, 2004. This situation is driving a realignment of markets and a restructuring of the industry that are well under way.

The response of textile manufacturers to intensifying global competition has been threefold. First, industry product mixes are shifting among countries. There is ongoing rationalization of production to emphasize products in which the manufacturer has the greatest competitive advantages and de-emphasize products that are more vulnerable to foreign competition.

In the United States and other developed countries, this implies a de-emphasis on textile products consumed in the apparel industry and a greater emphasis on other products, such as home furnishings and industrial textiles. Second, companies continue to invest in new, more productive plant and equipment. Textile production is no longer a low-technology, labour-intensive industry enterprise. Textile firms around the world are under constant pressure to become more efficient through technology upgrades. Third and perhaps most important, manufacturing is expanding into countries close to home markets. This is being undertaken primarily to support production sharing arrangements for apparel, sometimes in partnership with customers, suppliers, or other textile manufacturers. Textiles refer to yarns, threads and wools that can be spun, woven, tufted, tied and otherwise used to manufacture cloth. Production of textiles has been altered almost beyond recognition by mass-production and the introduction of modern manufacturing techniques.

MARKET STRUCTURE

Economies of scale in textile manufacturing are significant and create entry barriers in the market. The textile industry is more capital intensive than the clothing industry and is highly automated. It consists of spinning, weaving and finishing functions in integrated plants. For example, the approximate cost of a new fibre plant is $100 million depending on various factors. Costs of raw materials are volatile and can account for 50-60% of the cost of the finished products. For purposes of hedging against supply problems, many manufacturers are backward integrated into chemical intermediaries. Forward integration into apparel and product manufacturing is also seen in industry.

In 1995, the World Trade Organization (WTO), in the Agreement on Textiles and Clothing (ATC), agreed that all quotas on textiles and clothing would disappear between WTO member countries on January 1, 2005. The expiration of ATC marked the end of quotas, limiting textile and clothing trade between the WTO members. Developing countries such as China, India and Pakistan had been the most restricted by the quotas. While India and China are likely to emerge as winners, the main losers after quota will be quota-restricted countries that had enjoyed the benefits and protection for more than 40 years. The post quota market has changed with producers already affected by significant changes in retailing. Large retailers are now buying up independent brands to give consumers more value and enhance their shopping experience. Manufacturers in developed countries are more likely to adapt by relocating operations to production centers in low wage countries.

Some of the major textile industries can be divided as follows:

- Awnings, textile
- Blankets
- Bags or sacks, textile
- Blinds, textile
- Canvas goods
- Cordage
- Elasticized fabrics
- Fabrics, textile
- Felt (except floor coverings)
- Glass fibre fabrics
- Household linen
- Lace
- Narrow fabrics
- Netting
- Rope (except wire rope)
- Sailcloth
- Sewing thread
- Soft furnishings
- String
- Tarpaulins
- Tents
- Thread
- Towels
- Trimmings, textile
- Yarns

For U.S. textile producers, the most important regional expansion and partnering are taking place with Mexico. Broad market forces, including Asian pricing pressures, have accelerated the flow of textile capital into Mexico.

INDUSTRY DEFINITIONS

- *Acrylic*: Manufactured fibre derived from polyacrylonitrile. Its major properties include a soft, wool-like hand, machine washable and dryable, excellent colour retention. Solution-dyed versions have excellent resistance to sunlight and chlorine degradation.

- *Anti-dumping duty*: An extra duty imposed on an imported product by an importing country (or group of countries, as in the case of the EU) to compensate for the dumping of goods by a foreign supplier.
- *Aramid*: The generic name for a special group of synthetic fibres (aromatic polyamide) having high strength; examples are "Kevlar" and "Twaron".
- *Calico*: Tightly woven cotton type fabric with an all-over print, usually a small floral pattern on a contrasting background colour. Common end-uses include dresses, aprons, and quilts.
- *Countervailing duty*: An extra duty imposed on an imported product by an importing country (or group of countries, as in the case of the EU) to compensate for subsidies deemed to be illegal which are given to the manufacturer of the product in the exporting country.
- *Delocalization*: The geographical move of a production unit to a low cost country. (Note that the term is increasingly being used to describe all forms of shifts in production, including foreign sourcing and subcontracting.)
- *Industrial textiles*: A category of technical textiles used as part of an industrial process, or incorporated into final products.
- *Lycra*: DuPont's brand name for its elastane or spandex fibre.
- *PET*: Polyethylene terephthalate, the most common form of polyester.
- *Polyester*: A manufactured fibre introduced in the early 1950s, and is second only to cotton in worldwide use. Polyester has high strength (although somewhat lower than nylon), excellent resiliency, and high abrasion resistance. Low absorbency allows the fibre to dry quickly.
- *Quota*: A quantitative restraint imposed by an importing country on an exporting country - or established by agreement between the trading partners - which is designed to limit shipments of a product from the exporting to the importing country.
- *Spandex fibre*: The generic name used in the USA to denote elastane fibre. A manufactured elastomeric fibre that can be repeatedly stretched over 500% without breaking, and will still recover to its original length.
- *Technical textiles*: Textile materials and products manufactured primarily for their technical performance and functional properties rather than their aesthetic or decorative characteristics. End uses include aerospace, industrial, marine, medical, military, safety and transport textiles, and geotextiles.
- *Unit production systems*: An advanced apparel manufacturing system in which a single garment is progressed through a sequence of operations. Using a unit production system, a garment is automatically transported via a computer-controlled overhead hanging system,

which has been ergonomically designed to reduce the amount of handling of the garment (see also progressive bundle system).

- *Yarn*: A continuous strand of textile fibres created when a cluster of individual fibres are twisted together. These long yarns are used to create fabrics, either by knitting or weaving.
- *Yarn Numbering System*: Systems for sizing yarn fall into two basic types. The yarn number is based on the length of yarn needed to make up a specified weight. The finer the yarn, the higher the number. Cotton, wool and linen are numbered with such systems. The yarn number is based on the mass of a specified length of yarn. The finer the yarn, the lower the number. Silk, synthetic fibres and jute are numbered with such systems.
- *Textile HS codes*: Harmonized System Codes (HS Codes) are a government, regulated system used to classify products and their corresponding tariffs. The Harmonized System is an internationally developed and implemented commodity-description and coding system, on which the tariffs of most countries of the world (including the NAFTA countries) are based.
- *Manufactured Fibres*: It is important to understand that all manufactured fibres are not alike. Each fibre has a unique composition and physical properties. The U. S. Federal Trade Commission has established generic names and definitions for manufactured fibres, including acetate, acrylic, lyocell, modacrylic, nylon, polyester, polypropylene (olefin), rayon, and spandex.

SUSTAINABILITY

Sustain means "to maintain" or " to uphold" and with regard to industrial processes sustainability means establishing hose principles and practices which can help to maintain the equilibrium of nature in other words to avoid causing irreversible damage to the earth's natural resources. Moving to a greater degree of sustainability in our industrial processes and systems requires that we achieve a better balance between the social, economic and environmental aspects of textile production.

- A sustainable product is one that is manufactured in a way that respects the social elements of fair trade and human rights of the people involved in the whole of the manufacturing chain.
- A sustainable product is one that is manufactured in such a way that it has the lowest possible adverse effect on the environment *e.g.* by making the most efficient use of resources such as water and energy, and which goes the extra mile to recover raw materials, *e.g.* by the recycling of as much water as possible or by recovering the heat from wastewater discharges.

- But equally important, a sustainable product is one which can compete effectively in the global marketplace against less sustainable products. *i.e.* which offers value benefits to the consumer, and where the economic returns from its success are fairly distributed back along the supply chain.

In order to achieve the above agenda, it is important to work at each and every aspect of textile production where natural resources, energy and chemicals are consumed and emissions to air, water and land arise.

SUSTAINABLE TEXTILE PROCESSING

As we have already noted the textile industry is one of the most polluting industry sectors. A vast range and quantity of chemicals is used at every stage and the after- effects in terms of wastewater treatment and air pollution are critical to manage. There is a strong need to establish more sustainable textile processing measures in the industry.

The main objective of these measures should be to minimize and eliminate the most harmful inputs and the most polluting outputs and to reduce the level of chemical residues left on the textile. But it would be wrong to focus solely on the chemical inputs and ignore the consumption of energy and water which are the primary impacts of the textile processing industry.

In summary a sustainable approach covers the following points:

- Minimum use of resources (water and energy)
- Minimum chemical consumption
- Minimum pollution load
- Toxic chemicals eliminated from supply chain
- Harmful chemical residues eliminated from final textile

But in order to minimize the usage, it is important to measure the inputs and in order to eliminate the most harmful chemicals, it is important to know and understand what is being used. Uncontrolled or unknown inputs lead to unmanaged use of resources and uncontrolled outputs. The measurement of the parameters shown below is essential to establish an understanding of the input-output balance of the textile processing operation.

The measurement and control of these inputs and outputs can lead to:

- Improved resource productivity
- Improved eco-efficiency
- Improved cost efficiency
- Improved customer satisfaction
- Improved brand reputation

Once the understanding about the basics of processes and chemicals involved with the supply chain is established and awareness about the inputs is there, then control over the output of the production process can be achieved.

COLOUR COMMUNICATION

Finally we should note another important element that is often overlooked in seeking to improve the sustainability of the textile and clothing supply chain and that is accurate and timely colour communication.

We will return to this subject later in the series but for now a few tips for improving colour communication:

- Communicate your colour accurately using physical and/or digital colour standards. Accurate communication of the colour you require will reduce lab dip rejection rates and result in better right first time bulk dyeing.
- Communicate electronically when feasible: avoid wasting time due to delayed communication
- Do not set unachievable colour standards- avoid wasting dye and chemicals in trying to achieve heavy depths or brilliant shades on certain fibres.

CONCLUSION

Sustainable design, informed selection of dyes and chemicals, accurate colour communication, and controlled coloration using Best Available Technology are key elements in raising the standard of sustainability in the textile supply chain.

CREATING A GLOBAL VISION FOR SUSTAINABLE TEXTILES

Certification, such as eco-labels, plays a major role in giving credible assurance to retailers and end consumers that products comply with standards based on social, ecological and environmental standards. Of the 309 eco-labels identified world wide, 41 cover textiles (Ecolabelling, 2008) and some 9000 textile and clothing manufacturing companies have been certified.

Organic Exchange Fibre Report (2008/09) estimated a 54% increase in *cultivation* of organic cotton from the previous year, but *production* of organic cotton only 0.959% of conventional cotton, *i.e.* the growth in eco-labelled textiles is not reflected in consumer demand, raising questions about the impact eco-labelled or 'sustainable' textiles.

A number of issues may impede the spread of eco-labelled textiles through the supply chain: costs and time required to achieve, use and renew the eco-label, recession and potential loss of competitive advantages. This paper will present the findings from in depth interviews examining the decision making around buying and sourcing of eco-labelled fibre, fabrics or textile products.

The seven companies located both in India and the UK, spanned the supply chain, from fibre to product: textile manufacturers, eco-parameter testing labs, Certification Company and retailer. The aim of the research was to understand and investigate the marketing strategies for sustainable textile products. Our goal was to understand how designers, manufacturers and retailers may collaborate to deliver eco-labelled textiles attractive to the end consumer and we conclude by reflecting on potential implications for the supply chain integration

ECO-LABELS

Eco-labelling is becoming a differentiating factor on a worldwide scale in retail markets for textile and apparel purchase. Consumers are becoming increasingly concerned with the adverse impacts of industrial pollution on the environment and their health, resulting mounting pressure on textile, fashion industry to adopt more eco-friendly, chemicals and manufacturing processes.

Environmental concerns raised by production systems have been recognized since the late 1960's and attempts to move towards more sustainable and environmentally friendly approaches have been through a range of regulatory measures from green taxes to strict bans.

One approach acquiring increasing importance is that of 'environmental labelling' or 'eco-labelling', which, according to Piotrowski and Kratz (2005) differ in that e*nvironmental labelling* is broad and covers a range of labels and declarations of environmental performance and focus on consumption rather than the production of a given product; *e.g.* recyclable material while *eco-labels* are a sub-group of environmental labelling and convey environmental information about a product to the consumer and communicate that the environmental impacts are reduced over the entire life cycle of a product without specifying the production practices.

In brief an ecolabel:

- Identifies the overall environmental preferences of a product;
- Provides information on environment related product qualities;
- Are tools for consumers to identify environmentally safe product;
- Enables manufacturers to use ecofriendly raw material and ingredients;
- Is an additional product quality which can be used as a marketing tool;
- Can be issued by private or public body;
- Causes less stress on the environment
- Enables to earn premium on products.

SIGNIFICANCE OF ECO-LABELS

The eco-label has a role in the Integrated Product Policy (IPP) which aims to minimise the environmental degradation caused by any of the phases of a product's life cycle (tangible or intangible, such as service), *e.g.* manufacture, development, use or disposal (European Commission, 2008). All phases of a product life cycle are examined with the objective of improving their environmental performance. This approach requires all participants in this process to be engaged: eg, designers, industry, marketers, retailers and consumers.

The US EPA (1994) defined the following five factors for measuring effectiveness of an eco-label, the first four of which serve to support the last:

1. Consumer awareness ®of labels
2. Consumer acceptance of labels (credibility and understanding)
3. Changes in consumer behaviour
4. Changes in manufacturer behaviour
5. Net environmental gains

TYPES OF ECO-LABELS

Eco-labels may be voluntary or mandatory. Mandatory labelling is always third party labelling (*i.e.* an independent body is required to attest to required standards having been achieved), voluntary programmers may be established by firms or business associations as well as third party. Currently, there are no eco-labels in textiles and clothing enforced by mandatory rules. Eco-labels are normally issued either by government supported or private enterprises once it has been proved that the product of the applicant has met the criteria.,

- *Government*: Blue Angel (Germany), Eco Mark (Japan), Environmental Choice (Canada), White Swan (Nordic Countries), EU, Eco-Mark (India), Green Label (Singapore)
- *Private*: Eco-tex, Oeko-Tex (textiles and clothing) (Germany). Green Seal (United States),

The criteria for granting eco-labels are mostly based on the "cradle-to-grave" approach, *i.e.* the life-cycle analysis of the product and assessment of its impact on the environment from processing of raw materials, production, distribution, consumption and maintenance, (*i.e.* washing, ironing, dry-cleaning) and finally disposal of the product.

A 'Cradle to Cradle' certification programme assesses the sustainability of product ingredients for human and environmental health, as well as their recyclability or compostability making it easier at the design stage to create ecologically-intelligent products through choosing materials that meet key sustainability criteria for material health and material reutilisation (Braungart and McDonough, 2008). Differences between various eco-labelling schemes confuse public understanding of eco-labels: some are based on detailed analysis of the environmental impacts while others analyse only certain stages of the life-cycle.

Voluntary labels are classified according to International Standards Organisation (ISO) (Baumann, 2007). ISO is the world's largest non-governmental organization that develops and publishes International Standards and is a network of the national standards institutes of 163 countries. There are now eco-labelling schemes both in developed and developing countries and the ISO has classified the existing environmental labels into three typologies -Type I, II and III, specifying preferential principles and procedures for each one of them (beyond the remit of this paper to detail).

Many other prominent international trade and environmental organisations deal with issues related to eco-labelling, *e.g.*: the United Nations, the World Trade Organisation through its International Trade Centre and Committee on Trade and Environment, the US Environmental Protection Agency, as well as the Organisation for Economic Co-operation and Development (Naumann, 2001).

DEVELOPING AN ECO-LABEL

This is complex and complicated but can be generalised into four broad phases:

1. Selection of a product category by a labelling board through suggestions from industry, environmentalist, consumers, and other interested parties. De Man *et al* (1997) proposed that four levels of actors function throughout the industry:
   - Primary economic–Production/consumption decision-makers (producers, importers, consumers).
   - Secondary economic–influence the decision making of primary actors.
   - Governmental and administrative–set the framework for the actors.
   - Others–try to influence the behaviour of all actors to improve status quo.
2. Life-cycle analysis to assess environmental impact of products in chosen category and examine the material and energy inputs for manufacture and use of a product and the solid, liquid, and gaseous waste generated at each stage of lifecycle, *e.g.* raw material, production, distribution, packing use and disposal.
3. Criteria and thresholds for the award of an eco-label set taking into consideration technical feasibility and environmental impacts in different media like air, water and soil against one another. Different eco-labels have differing methodologies, *e.g.*, Oeko-tex 100 examines harmful residues on the product, while GOTS tends to look environmental as well as residual parameters.
4. The product category and criteria is reviewed and refined. Interested parties including industry and environmental and consumer groups are asked for their inputs, although they are often already included much earlier on in the process.

ACQUIRING AN ECO-LABEL

Atilgan, (2007) identified fourteen steps to obtaining the EU Flower eco-label for a Turkish textile manufacturing firm In brief, the factory administration decides to apply for eco-label and use it. At the performance control stage, the selected ecolabelling organisation's prohibited chemicals are identified. Detailed monitoring of the chemicals and quality control system is necessary at each stage of production to prove that the products introduced to the certificating institute are appropriate.

At this stage, honesty is necessary because the institute is authorised to make tests any time it likes. Failing in these tests results is cancellation of the eco-label certificate (Atilgan, 2007). As well as the time and complexity of obtaining the eco-label, the most significant issue is that of cost, which, for the EU eco-label is set at 0.15% of the annual turnover of the eco-labelled product, costing up to □1,300 for registration (*i.e.* to apply for the label), □25,000 per year for the use of the label, with a reduction of 25% for SMEs (buyusa.gov 2009, Rubik and Frankl, 2005).

Atilgan (2007) indicated that the costs of using eco-labelled production made the finished product between 12-15% more expensive to make, depressing interest in their use by manufacturers and retailers. The costs become even greater down the chain. A UK corporate wear supplier estimated the costs of using eco-labelled fabrics (such as Teijin fabrics, from EcoCircle closed loop system of fibre processing) as placing a 57% premium on their final product (Sinha and Hussey, 2009).

ECO-LABELS AND CERTIFICATION

Regardless of the costs/benefits, Atilgan (2007) urged the Turkish government and industry to become engaged with eco-labelling as he felt that the next area of purchase and trading selection appeared to be based on the criteria set by the eco-labelling bodies. Furthermore he suggested that these costs may be mitigated through the use of smaller amounts of high quality products, optimising the production techniques *e.g.* by controlling all recipes and procedures, and identifying problem areas.

Given the premiums on using eco-labelled textiles, commercial buyers of such products require assurances that goods comply with standards set by the eco-label organisation. This is particularly relevant for organic fibres where the market has flooded by products described in vague terms such as 'Green' and 'Eco' (Rundgren, 1999). To differentiate between sustainable and traditional textiles, third party standards are desirable.

The unverified, market-based self-labelling (without outside monitoring) approach for textile ecolabelling, invites fraud due to lack of third-party verification (Moore *et al.*, 2009). Retailers and consumers are starting to demand labels backed by solid third-party certifications to give confidence to all the marketing claims. Commercial buyers therefore may require certification of the product. Like ISO Type I labels, certification schemes are voluntary and provide information on the environmental impacts of a company's production methods and processes (PPMs), ie, impacts of the *entire* activity not just a particular *product* (Rotherham, 1999). Certification provides a comprehensive system for ensuring that certain standards of organic production and processing are met.

The system includes:

- Developing rules or standards (standard setting).
- Verifying and evaluating performance against those standards (inspection).
- Recognizing procedures which successfully meet the standards (certification).

SCOPE AND TRANSACTION CERTIFICATES

Most companies only look apply for certification if it is required by the customer. From the literature reviewed to date, shows the flow of the certification across the supply chain. There are two kinds of certificates: Scope - issued to the selling company and stating its name, the products and production facility inspected and certified in accordance with standards. Transaction: only issued (to the seller) for the sale of products if the applicant has the scope certificate.

For example, a retail buyer may require a scope certificate from chosen garment manufacturer to prove that their manufacturing facility is certified and products are being produced in accordance with standards. If the manufacturer is not certified, they contact the certification body for the certification. If the garment manufacturer does not manufacture textiles, then the retail buyer would need to source the required certified raw material to process into fabric. Then they will contact the factories from where they wish to buy the fabric, however that facility must also be certified and to prove it, processor need to provide scope certificate to garment manufacturer, if there is no certificate then they need to contact the certification body to certify their facility.

If the ordered garment has embroidery and printing, then either the garment manufacturer should acquire certification for those facilities under their scope, or suggest to the embroiderer or printer to acquire certification themselves through using certified threads, dyes or chemicals.

To obtain a certificate can take between 30-45 days, depending on the standards, factory condition, number of factories under one application understanding of standards by applicant, and changes within the factory with respect to compliance with standards. This process continues 'backwards' along the supply chain till supplying of certified (in this case organic) cotton to spinner.

CERTIFICATION FLOW

If all the points are certified the process would take around a minimum of 30-45 working days to issue a certificate from certification body. If during inspection any non-compliance occurs the process would be lengthened as each certification takes about 30-45 days. A further complication is that the entire chain requires consistency in certification bodies used: *i.e.*, if the garment manufacturer is required to produce to GOTS standards, the processors must also have GOTS certification-a different certification will not suffice as they each have their own analysis processes for certification.

In brief, certification is a complex, time consuming and costly process however it creates the transparency within supply chain. For each step in the supply of a product for the chain, the transaction certificate provides transparency within the supply chain and it gives assurance to buyer about the compliance of products with standards. The certification body issues the transaction certificate for each application made by their certified clients and is delivered to the buyer to give assurance and to confirm the product is manufactured accordance with standards.

- Demand for Factory certification
- Receiving certified product from supplier
- Contacting certification body for certification
- Contacting ecological testing laboratory for required test reports

COST AND BENEFITS OF CERTIFICATION

An important factor in the success of any eco-labels is its ability to cover its certification costs and therefore stay in business.

According to EPA (1998), the ease with which programs will be able to cover costs varies depending on two questions:

- Can the programme charge enough in application, testing, audit and other fees to cover its costs; and
- Can the programme subsidize its environmental labelling activities from other programme activities?

Cost of certification can be high in relation to the value of the product and thus can become prohibitive. This is especially true for textiles because of the number of process are involved from production to consumer. It may be more expensive for companies in developing countries to obtain labels and certifications, "due to factors such as the lack of existing management structures (EMS), the novelty of EMS, insufficient infrastructure, and high auditing costs if companies have to rely on international consultants and certification companies" (Rotherham, 1999).

In addition to capital costs, the absence of necessary knowledge and skills and a lack of mutual recognition between different national programs can further disadvantage some countries (UNCTAD 1997).

ECO-LABELLED SUSTAINABLE TEXTILE PRODUCTS:

'Sustainability' is the ability to maintain an activity indefinitely over time. A sustainable activity is therefore one that does not exhaust the resources on which it depends. The concept of sustainability gained worldwide recognition following a report in 1987 by the World Commission on Environment and Development (WCED). This report entitled *Our Common Future*, defined sustainability development as (Performance Apparel Market, 2009) –

"Development that meets the needs of the present without compromising the ability of future generations to meet their own needs."

Based on above definition, it's possible to define the term 'sustainable textile product' (STP) as –

"a product which is manufactured with the help of services and related products in responds to basic need and bring a better quality of the life with use of minimum amount of natural resources and toxic chemicals during the production, minimum waste emission to the environment over the life cycle of product keeping environmental and social factors in mind throughout the supply chain".

Sustainable textile or apparel are, therefore:

- Safe for human and physical environment;
- Made from renewable materials;
- Produced while making the most efficient use of resources such as water and energy;
- Manufactured by people employed in decent working environment;
- Capable of being washed at low temperature using environmentally friendly laundering agents; and
- Capable of being returned safely to the environment at the end of their useful life (Performance Apparel Market, 2009).

CHALLENGES FOR ECO-LABELLING

There are many challenges for the eco-labelling, the most serious of which are: misleading or fraudulent to uninformative claims, unfair competition and protectionism and lack of stringency or standardisation in the process or mechanisms of eco-labelling.

The objective of certification is to gain access to the market for environmentally sustainable products (Rundgren, 1999), and the certification process should help as data collected in the process of certification can be very useful for market planning as well as for extension and research, moreover, improves the 'image of product and increases its credibility and visibility (Rundgren, 1999), Auriol and Schilizzi's (2003) studies have shown that the costlier the certification process, the fewer firms able to afford certification, ie, cost becomes a major factor in deciding market structure, potentially leading to monopoly and ultimately to no certification at all. Influencing factors on consumers' willingness to buy environmentally friendly products have been identified and categorised as: demographics, knowledge, values, attitudes and behaviour (Laroche *et al.*, 2001). However, price has been found to be one of the most decisive factors in determining when consumers actually purchase apparel products.

Consumers' willingness to pay and purchase cloths made from sustainable raw material like organic cotton is a complex issue (Gam *et al.*, 2010). Empirical testing has shown, however, that many consumers are willing to pay a premium for eco-labeled products (Imkamp, 2000; Loureiro *et al.*, 2002; Makatouni, 2002; Moon *et al.*, 2002) and that they do purchase such products (Lathrop and Centner, 1998; Teisl *et al.*, 2002). According to Gam *et al.*, (2010) study in one of US, only 35% (27 out of 84) were willing to pay more for OCC and only 10.7% (9 out of 84) were willing to accept more than a 10% increase in price for OCC. In contrast, they found that 52% of survey participants would pay a 50% price premium and 25% of the participants would pay 100% more for an OCC over conventional cotton clothing.

From above results it's clear that, willingness to pay for STP like product made from organic cotton varies place to place, country to country. Differences in testing and certification methods have created difficulties in the application of an eco-label to a particular product category. For example, should the label represent an overall assessment of a product's environmental burden over its entire life cycle, or some subset of it? What techniques can be used to measure environmental impact? Who determines what specific environmental impacts are the most important? And what criteria are appropriate in rating impacts? Moreover, the consumer is unable to verify the claims made by the eco-label.

An analysis of ecological labelling process by Lavallee and Plouffe (2004) concluded that 'cradle-tograve' analysis for ecolabeled products and services is not always, in fact, respected, and that at the present time ecolabel delivery criteria are not sufficiently stringent or standardised leading to confusion in the marketplace, making it difficult for companies to identify stakeholder preferences and for justified environmental claims to be considered credible (Rotherham, 1999; EPA 1998).

THE RESEARCH QUESTION AND METHODOLOGY:

The research undertook to understand the issues within and across the textile supply chain that come to bear upon the growth of eco-labelled sustainable textiles products. While the fear of losing market share is a motivating factor, it should be stressed that market impact of eco-labelled textiles products are only one indication of an ecolabelling programme's success.

The effectiveness of an eco-label ultimately depends on the extent to which consumers perceive, recognise and act on the information it conveys. The next section describes the research methodology adopted and presents some of the most pertinent results. Multiple in-depth interviews were conducted using semi-structured interview schedules designed as open ended questions. The decision was made to focus the study on organic cotton (one of a range of sustainable fibres) as this (and its products) is produced in large quantity, consumers are much more aware of it, retailers as well as consumers have accepted it and year by year demand is growing throughout world. 9 in-depth interviews were conducted with company employees who ranged from director of development, director of Marketing, Company Owner/Director, etc. As indicated, the companies were selected on the basis that they had at least one following production facility in order to complete as much as possible the supply chain: spinning, weaving, knitting, wet processing, stitching/garmenting, retailing, testing, authorization of certification and authorization to issue eco-label.

Companies interviewed are listed in table 1; the majority of the companies were in India as this is a centre of organics cotton production.

RESULTS

As mentioned earlier, the research tried to understand the issues within and across the textile supply chain that come to bear upon the growth of eco-labelled sustainable textiles products in particular and in developing a sustainable textiles industry in general. From the interviews, the following issues arose:

SUSTAINABLE TEXTILES PRODUCTS (STP) ARE NEEDED

All companies interviewed agreed that STPs are needed. According to textile manufactures companies A and B, STP is a holistic approach and it can be achieved through recycle, reduce and reuse processes. They felt it very important to note that all naturally grown products are not organic or sustainable; for example, organic cotton. All naturally grown cotton is not organic; it might be genetically modified organic cotton. Also, there is no assurance that the land doesn't have any traces of harmful fertiliser, pesticides. According to dyes and chemicals manufacturers companies D and F, STP is a mindset (company D) and its "the product which has manufactured by taking care of all the three elements of sustainability that is: social, economic and environmental sustainability and product design is fashionable to sell and sustain in the market" (company F). STP's are needed to take care of our eco-system, so that this planet can 'sustain' the lives and livelihood 6 billion inhabitants.

From certifiers (company G), STP's are those products which are manufactured from fibre cultivated by natural or organic method considering social and environmental impact, understanding soil fertility and animal welfare. Further, Mr A (Company H) added that "producing sustainable textile is one of the ways to overcome the Global warming" and Ms. AB (company I) suggested product made from 'organic cotton' are the best example of STP.

'WHY' AND 'HOW' TO BECOME SUSTAINABLE TEXTILE MANUFACTURER

A number of methods are available to enter the sustainability arena; those raised by the interviewed companies included the following:

TRANSPARENCY AND EXTENT TO WHICH THE COMPANIES HAD TAKEN UP ECO-LABELLING

All the companies interviewed were either certified or were in the process of receiving certification, ie there was felt to be a general endorsement of the idea of eco-labelling.

BOLSTER THE CLAIMS WITH INDEPENDENT VERIFICATION

Third-party verification of environmental credentials can often bring legitimacy to sustainability. Many of the most successful eco-labels are those that have been backed by issues-led organisations, for example GOTS certification for textile products made from organic cotton. Third-party verification can range in scope from qualitative assurance of general claims to detailed verification of all stages of a full life-cycle product assessment.

Given the generally low levels of consumer trust in big business, some degree of external verification is an essential component of any credible environmental claim. According to Mr. H (company G) and Mr. A (company H), the claims made by the manufacturer or retailers are cross checked by certification of responsible eco-labelling body by testing the goods which can be picked from market store to confirm whether claims are right or wrong.

In case of dyes and chemical, customer requires proof from third party initiating business. Therefore many companies use the test reports as a marketing tool to prove the product and company integrity. Within this study, it was the analysis, researcher has observed that, all the segments of supply chain apply for certification or verification as a buyer requirement.

EDUCATE, ENABLE, AND ENCOURAGE

Educating, enabling, and encouraging people to act towards sustainability is key for the success of any eco-label and STP as consumers' usage and disposal patterns liberate CO_2 and so there should be programmes to educate them. Methods to do this range from placing trust worthy eco-label with required info on it, through various media, and through regulations.

Once the consumer gets educated then may be encouraged to prefer buying more sustainable products by linking them with promotions and reward schemes (carefully and consistently in accordance with principles of sustainable consumption). Product information can now be shared via many more 'touch points': at point of sale; in retailer magazines, leaflets and websites; through road shows, help lines and education packs. Enlisting employees to promote sustainability is another method as employees are key players of any manufacturing facility. Investing in the skills of employees is part of a sustainable and responsible human resource management (Brito *et al.*, 2008). To educate their employees, company A celebrates days like Green Day, Earth day in factory to increase the awareness of sustainability.

Company B have organised seminars for their employees, their suppliers, buyers, contractors etc before starting production of organic cotton where they have invited expertise from industry on subjects like eco-friendly dyes and chemicals, GOTS, ethical production practices etc.

REDUCE, REUSE, AND RECYCLE:

To become sustainable, everyone, from manufacturer to consumer should think of 'Reduce', 'Reuse' and 'Recycle'. Textile waste in landfill contributes to the formation of leachate as it decomposes, (which has the potential to contaminate groundwater), methane gas (a major cause of greenhouse gases contributing to global warming) and ammonia (highly toxic for land, water and air) (Productivity Commission 2006).

The companies interviewed engaged currently in the following activities to try to reduce waste and reuse material:

Company A:

- By utilizing solar energy company is running around 20 computers within factory,
- From cutting waste, manufactured 5000 bags and sold at local super market subsidiary price
- Currently working on yarn manufactured from solar energy,
- Reusing treated effluent water for gardening and washing purpose (specifically washing of printing screens).

Company B:

- Have reverse osmosis (RO) plant for purification of processing effluent, after purification of that water it utilised for washing, gardening etc.
- Have started recycling of old paper cones (used in spinning for packing) and making new ones.

Company I:

- Offering shopping bags to their customers made of potato starch *i.e.* 100% biodegradable.

DISCUSSION AND CONCLUSION

The proliferation of voluntary certification and labelling schemes for environmentally and socially responsible production is often seen as a driven by companies and consumers. Consumers are heavily involved in environmental pollution because of their buying behaviour and consumption of textiles and significant associations were found between environmental shopping attitudes and behaviour and willingness to pay more for organic cotton products and observed that consumers with a greater environmental awareness demand more environmentally friendly merchandise (Fraj and Martinez, 2006).

Eco-labels are not simple to understand as they appear and so may not be as appropriate marketing tools as suggested by the government policies. For example, GOTS and OE standards labelling guide are both standards applicable for products made from organic cotton on which retailers, manufacturers can use respective logo on their tags.

Under both the standards it's mandatory to mention the percentage of organic cotton on the label and if the product is made from 100% organically cotton, manufacturer or retailer can use the statement "Organic" and "Made with 100% organically grown cotton" respectively. However, GOTS is based on social, technical and environmental areas while Organic Exchange 100 Standard (OE 100) is for tracking and documenting the purchase, handling, and use of 100% certified organically farmed cotton fibres (or organicinconversion cotton fibre) in yarns, fabrics and finished goods.

The significance of this difference that, while purchasing the product a consumer will check only the organic content and assume that product is eco-friendly or sustainable. However, this is not a completely accurate picture as an OE standard does not look into social or harmful dyes and chemicals or about the environment related issues and an OE 100 logoed garment may be made from 100% organically grown cotton but then finished with harmful dyes and chemicals, printed with non-eco-friendly printing technique like solvent based printing.

Therefore a large problem in marketing communication is using eco-labels is the lack of common definition or general understanding exists for what constitutes environmentally friendly clothing and eco-labelling. Eco-labels backed by solid third party certification give confidence to consumers about the genuineness of product. Certification bodies are the key player for the growth of eco-labelled STP who confirms the product, process and manufactures integrity with respect to sustainability. In this study, across the whole supply chain, most of the people had positive thought about requirement of certification.

However, cost of certification can be high in relation to the value of the product and thus can become prohibitive. This is especially true for textiles because of the number of process are involved from production to consumer. To overcome the certification and labelling cost, Ibanez and Grolleau (2007) suggested "carrot" approach - the firm who preserve the environment appreciate them by subsidizing a recognized labelling and "stick" approach - increase of labelling cost for polluting firms by enforcing stricter labelling guidelines and severe punishment in case of deceptive use of environmental claims.

According to Getz and Shreck (2006), despite much analysis of third-party certification, little is known about how certification is enabled or enacted at the point of production. The insights of those few who have explored some of the political and social effects of certification at the point of production are worthy of further examination. In this study researcher has analysed the 'how' certification process works within the organic textile-clothing industry and 'why' it is required. Having reviewed the process of certification and how individual actors of supply chain get involved into this very complex, time consuming and sometime costly process of certification. Therefore we suggest that manufacturers take a vertically integrated approach through networking or developing two or three processing facilities to overcome cost and time of certification, reduce handling, transportation CO2 emission and time to retailers' shelves. Companies such as M&S, H&M, Zara, who are bringing STP's to the mass market must be operating in this manner.

Moreover we suggest, a fourth to the traditional three pillars of sustainable development; "technical" by which the first three may be achieved.

- Demand for Factory certification
- Receiving certified product from supplier
- Contacting certification body for certification
- Contacting ecological testing laboratory for required test reports

SD pillars Why How

- Economic Competitiveness
- Process and product innovation; Process and product substitution.
- Environmental Clean out puts Reduce, Reuse and Recycle
- Social Social fairness Better human resource management
- Technical Eco-friendly inputs Use of certified products

LIMITATIONS TO THE RESEARCH

This study's limitations need to be taken into account when considering its contributions to theory and findings. This being a one year period of study, time and cost were main limitations and affected the sampling, since it was impossible to visit entire population due to budget restrictions, large dispersal of locations.

It was also not possible to include either the logistics sector (also responsible for CO2 emission); government authorities' whose views may put some light on government policies and approach towards eco-labelled STP or the manufacturers who are not certified by any of the standards to understand why they are not willing to get certify.

The results have not been validated, again due to lack of time. Bias may arisen, some participants tend to express views that are consistent with organic cotton, GOTS, OE standards only and try not to present themselves negatively. In addition, the interviewees may have also been unable to disclose some information due to the privacy issues with various companies. Information collected may be prone to some inaccuracy as a result of less than accurate recall, lack of information, or discomfort with self-disclosure.

MARKET METRICS

UNITED STATES

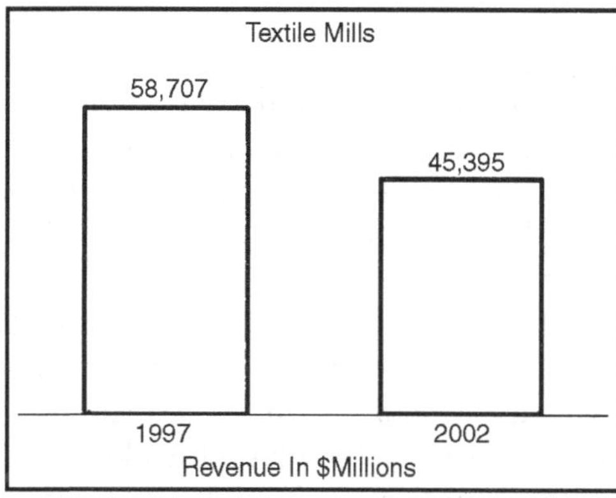

The textile industry is currently more than US $400 billion and continues to expand. The globalization of the textile trade has increased the outsourcing of production.

A variety of fabrics are used worldwide in different applications such as apparel, household textiles and furnishings, medical equipment, industrial and technical products. Recent studies have highlighted that fabric weaving alone consumes around 28 million tons of fibre annually. It is predicted that global production will grow by 25% between 2002 and 2010, to reach more than 35 million tons and Asia is one of the key regions for growth.

The United States ships raw materials to other countries in which the products are made. The cost of labour is cheaper and provides retail companies the ability to buy and sell at lower costs. This increases competitiveness and has led to several textile companies having to shut down operations.

At present, there is a minor trade war developing between the United States and China in regards to textiles and apparel. Chinese exports of textiles to the United States increased by 320 per cent between January 2002 and November 2003 while prices of Chinese textiles and apparel declined by 55 per cent. The growth in imports of textiles to U.S. during 1995 to 2002 was about 9% annually in dollar terms.

INDUSTRY PLAYERS

These are major players in this market, but not an exhaustive list of all key firms. Revenues, Net Income and Market Capitalization are expressed in US$ Millions.

Ticker	Company	Location	Revenue	Net Income Margin%	Profit Cap	Market
VFC	V F CP	Greensboro, North Carolina	6860	590	7.81%	8040
LIZ	LIZ Claiborne	New York, New York	4950	149	2.75%	2410
HBI	Hanesbrands	Winston-Salem, North Carolina	2250	74	3.28%	2770
RL	Ralph Lauren Polo CP	New York, New York	4540	387	8.50%	6870
ZQK	Quiksilver	Huntington Beach, Califormia	2362	93	3.94%	1280

Research and development and design are important in industrial textiles where the material technology provides competitive advantage. In industrial countries, notably the United States, an increasing share of the textile sector produces household appliances and other industrial fabrics such as for the furniture and car industries.

This is a more R&D intensive segment of the industry and subject to less frequent changes in patterns, material and colours. Only about a third of US textile production was used for clothing in the late 1990s.

- Mohawk Industries, Inc., is the market leader both in terms of market capitalization and annual sales in 2006. It engages in the production and sale of floor covering products for residential and commercial applications in the United States and Europe. It operates in three segments: Mohawk, Dal-Tile, and Unilin. The Mohawk segment designs, manufactures, and distributes floor covering products, including carpets, rugs, ceramic tiles, hardwood, resilient, and laminates. It markets its products through independent floor covering retailers, home centers, mass merchandisers, department stores, commercial dealers, and commercial end users, as well as through private labeling programs.
- Albany International Corp. produces paper machine clothing and doors worldwide. It operates in three segments: Paper Machine Clothing, Applied Technologies, and Albany Door Systems. Paper Machine Clothing segment includes fabrics and belts used in the manufacture of paper and paperboard. This segment offers forming, pressing, and dryer fabrics, and process belts.
- Polymer Group, Inc. manufactures and markets nonwoven and oriented polyolefin products. Its Nonwovens division provides nonwoven materials, which are used in diapers, training pants, feminine sanitary protection, adult incontinence, baby wipes, and consumer wiping products. Its products are used in medical applications, such as wound care sponges and dressings, and disposable surgical packs, as well as for apparel, including operating room gowns and drapes, face masks, and shoe covers; and in industrial applications.

- Interface, Inc. engages in the design, production, and sale of modular carpet products in Americas, Europe, and Asia-Pacific. It also manufactures broadloom carpet, panel fabrics, and upholstery fabrics for the commercial interiors industry.
- International Textile Group, Inc., together with its subsidiaries, operates as a textile manufacturer. The company produces automotive airbag fabrics and airbag cushions for automotive airbag modules. International Textile Group also manufactures and sells various technical and value added fabrics for use in fire service apparel, ballistics materials, filtration, military fabrics, and outdoor awnings and covers. The company also produces denim, cotton, synthetic and worsted apparel fabrics for use in bottom garments, such as pants, skirts, and shorts. It manufactures fabrics for military dress uniforms and battle dress uniforms for the U.S. Government and government contractors.
- Xerium Technologies, Inc. manufactures and supplies consumable products used in the production of paper clothing and roll covers.
- The Dixie Group, Inc. engages in the manufacture, marketing, and sale of carpets and rugs to residential and commercial customers in the United States.

TRENDS AND RECENT DEVELOPMENTS

To sharpen their competitive edge, many U.S. textile companies are,

- Restructuring their operations through mergers, acquisitions, and divestitures,
- Concentrating on more profitable niche markets and product diversification,
- Improving productivity through new capital investment,
- Building strategic partnerships with retail and apparel companies,
- Using quick response and other communication-oriented techniques to enhance their overall flexibility and competitiveness.

U.S. textile companies have been acting to improve their competitive positions through reorganization, restructuring, and downsizing. By merging, textile companies gain market share and access to capital. Through acquisitions and vertical integration, textile companies have been able to achieve economies of scale and expand the range of their services.

Another trend is for U.S. mills and converters to move slowly away from staple fabrics for the apparel industry and toward niche products and diversification into multiple sewn products markets. Technical fibres for sportswear and active-wear have experienced tremendous growth. Other profitable growth opportunities in non-apparel fabric markets include automotive, medical, fire safety, and telecommunications (fibre-optics).

The U.S. textile industry is a technologically advanced industry, achieving such advancement through restructuring and investment in modern machinery. Spinning, weaving, and knitting processes are highly modernized in many companies, and this has allowed them to reduce the share of labour in production costs and increase quality and speed. There has been a substantial job loss in the U.S. textile industry, a trend that has been ongoing for almost 30 years. This job loss can be attributed partly to imports, but like many other industries in the United States, the textile industry has been able to increase its production through new technologies while decreasing the number of employees.

Overall textile production has increased more than 20 per cent in fifteen years even though industry employment has dropped over 13 per cent. The textile industry continues to invest in machinery and technology, spending upward of $2 billion a year to maintain modern manufacturing facilities. Continued investment in modern technology, including information technology, is very important. Textile production is very capital intensive, and up-to-date technology is essential to meet the increasingly rigorous demand for high-quality products.

Proximity to major markets has assumed increasing economic significance and tariffs are restraining trade due to the fact that products cross borders several times. Developing countries are catching up with China in terms of unit labour costs in the textile and clothing sector. Future trends include multifunctional textiles, intelligent textiles, eco-textiles, e-textiles and customised textiles in the textile-apparel industry. Nano-Tex is the company manufacturing "enhanced textiles" which are stain resistant transparent fabrics manufactured using nanotechnology.

5

Manufacturing Process in Textile Industry

In the western countries particularly, the stress is being laid lower for changes in technology as faras its process is concerned Textile, fiber and apparel industries are mainly such industries. The causes why technological change is currently receiving so much attention are the same for nearly all of our manufacturing industries, but they are especially emphatic for these textile industries. The most important of these, as we all know, is import competition.

Only 10 or 15 years ago, imports did not appear to be much of 6 threat to the United States textile industries, but since that time they have increased at ever greater rates to the point where, in some categories of apparel, some 50-60 per cent of the goods that Americans consume are imported. In recent past, an increase in imports did not appear quite so important, since our own industries were managing to keep roughly abreast with exports. However, in the last five years, that situation has changed and today the imbalance between imports and exports is huge.

The chief textile industry has been suffering its own import problem and textile manufacturers have, correctly, lamented what seems to be an import situation going out of control. However, the data display that this situation is compounded several-fold for the apparel industry. The problem is made even more severe by the fact that the United States markets for textiles and apparel have reached a plateau.

Over the last 15 years or so, the average yearly consumption of textiles in all forms has peaked at something in the neighbourhood of 60 pounds per person per year. Although there are substantial variations about that figure from year to year the average has remained more or less constant.

This implies that domestic and imported textile items are competing for the same markets and are not competing in an expanding market. This situation has led to a greater, or perhaps renewed, stress on "productivity" in the textile manufacturing industries. It is thought that increasing productivity will somehow make domestic products more competitive with imported products, which are widely thought to be competitive only due to low labour rates. Sometimes this is true; other times it is not.

The import situation also has brought about a renewed discovery that all facets of the textile industries are quite closely interrelated. It is widely understood now that factors that hurt the apparel industry's ability to be competitive also hurt the fiber and primary textile industries as a chief supplier. This chain also comprises materials, i.e., polymer and fiber producers on the front end, and retail operations on the back end. Fiber producers are also very aware of the import problems of the apparel industry, but an interesting dichotomy seems in the retail end of the business, in which it sometimes appears that the importation of relatively inexpensive apparel is actively encouraged.

What is this technological change on which we collectively pin our hopes for the future? A general term that we hear often is automation. Automation can be classified in two ways: by its technical characteristics and by its effect on the industry. Generally, automation can be characterized according to one of three technological types. The first is mechanization. The term has found uses in most manufacturing industries, but the apparel definition is of particular interest here.

Mechanization means, in this context, the building and utilisation of devices that can be described as worker aids, i.e., rather simple mechanical devices generally resemble to templates, guides, or other positioning devices that help the operator move material reliably or perform simple operations on that material. Instances in the apparel industry would comprise a variety of guides to bring fabric to the sewing machine head and, similarly, simple devices to fold fabric' reproducibly for certain kinds of seam formation. The second kind of automation with which we should be very familiar in the textile industry is hard automation.

Loosely defined, hard automation is identified with machines that are constructed to do one specialist job very well. The negative connotation of hard automation is that these machines can do only that one job; to do any other they must be completely torn apart and rebuilt. Instances comprise many varieties of modern yarn spinning machines, knitting machines, and looms.

The third automation classification is intelligent automation; this is the kind that we hear most about today. Its most advanced form perhaps is the robot, and a rohot definition shows exactly what intelligent automation is supposed to be about. The key concepts in such a definition are flexible, reprogrammable, capable of doing a number of tasks, and capable of change to do a number of other tasks. Automation also can be classified by the effect it has on the industry itself.

In this perspective, we can distinguish between "evolutionary" effects and "revolutionary" effects. Evolutionary effects are those in which a machine effectively replaces a worker on a one to one ground. The machine may perform its job somewhat better than the worker or not, but the key concept is that what is being done is little different than it was before. A nearly perfect instance is the technology involved in automatic knot tying in the textile industry. This in is not quite so good if we consider the emergence of yarn splicing as a substitute for tying; no human could ever splice in the way that modern splicing machines do.

On the other hand, textile technology is full of examples of revolutionary automation—the invention of machines that indeed did revolutionize the way in which textiles were produced. Some instances from yarn technology are particularly appropriate. Using the pre-Industrial Revolution spinning wheel as the beginning point, the invention at the Industrial Revolution of a succession of various ways of making staple yarn—the spinning mule, the spinning jenny, the spindle and flyer mechanism-all of these led to entirely new products; therefore, these machines must be considered revolutionary.

The modern day descendants of these machines are the open end spinning machine and the many friction twisting machines, which again operate so differently from their predecessors that they must be considered to have revolutionized the yarn-making business.

To understand what is going on in automation today, it is fruitful to consider what has happened in the past in the introduction of automation in a number of industries. As a first instance, let us consider hard goods manufacturing, which we can consider to be the making of automobiles, refrigerators, or even electronic components.

From the days in which these devices were assembled completely by hand by master craftsmen, to Henry Ford's assembly line concepts, to today's use of the modern assembly cell in which one combines hard automation with robotics, computer control and computer aided design, hard goods manufacturing has progressed steadily through the way of technological change. We should all be related to some of the same progression in the manufacturing of textile products, From the earliest times to the English Industrial Revolution textiles were produced almost completely by hand.

The only concession to machinery that one might detect was in the area of mechanization or worker aids. Even just pre-Industrial Revolution, for example as textiles were made in the cottage industry of the United States before the Revolutionary War, whether we are talking about carding, or spinning, or weaving, the equipment utilised in that era was composed inevitably of worker aids, which were used to orient and maintain the position of the parts while a human assembled them into a textile product.

However, after the Industrial Revolution and up to the present, substantial and continuous progress in the hard automation era has taken place. The last ITMA machinery in Milan showed a number of examples of new hard automation; one of the most striking, probably, was a device to automatically and continuously transfer ring-spun packages to a winder, highly automated in the classical hard automation sense. As an instance of intelligent automation, we can consider, besides robots, automatic guided vehicles which are seen in increasing numbers in textile plants today.

These are flexible as they are computer controlled, and they do what they are told when they are told to do it. While the tractor part is completely flexible the specialized trailer that it tows is an instance of hard automation. The out come of all of this activity over 200 + years has been that the number of man-hours needed to make a 'Unit amount of a textile product, for example staple yarn or woven fabric, has decreased by a factor of 10 every 50 or 60 years since 1760.

That represents a gradual increase in productivity brought about by the fact that new inventions that drastically improve productivity were introduced periodically into the industry. The technologies that appear to hold the most hope for continuing this trend today all are related to computers, micro-processors, robotics, and other forms of flexible automation.

Progress in the apparel manufacturing area seems not to have been nearly so gradual. There are few studies in this area and it seems to the casual observer that little progress in manufacturing technology has been made since Elias Howe's invention of the sewing machine in 1846. Several present day apparel plants still rely almost totally on human labour, have confusing paths of material flow, and are operated on a piecework basis.

On the other hand, a few plants have a high level of mechanization and hard automation, particularly where great quantities of one product are manufactured, for instance, denim jeans. Dramatically improving the productivity of the apparel industry may well require use of the same concepts that are important for the textile industry—robots, materials handling improvements, vision systems—a whole variety of technologies that are feature of those being investigated now in the "Draper" project of advanced automation. There explicity are reasons to adopt technological change, but they are not always the obvious ones.

The reasons usually given fall into three general categories: the reduction of labour costs, the increase of quality of product, and the increase of flexibility of process. When most of us think about altering labour cost by adopting machinery, we think regarding replacing workers with machines. And, in fact, this is very often the case. But it also is generally the case that this is not the main source of labour savings.

The automobile industry, for instance, found out a number of years ago that one of the most compelling reasons for embracing robots for their manufacturing processes and replacing human workers with that technology was not in the current hourly "wage" comparison but rather in the future effects that inflation had on workers salaries versus the future cost of keeping the robot in operation. As another instance, the automobile industry has also found that labour costs are effectively reduced simply by virtue of the fact that adequately chosen automation can do a job more efficiently. In the automobile industry, a common instance is spray painting, in which the chief cost savings is the cost of the paint that is wasted. These really are indirect labour costs. A third indirect labour cost is present in the argument that machines can do jobs that people no longer want to do, or that they can do jobs on a regular basis for which there is a fast human labour turnover; and therefore very large associated training costs.

Automation can also increase the quality of a product and there are many examples in the textile industry of hard automation that does exactly that. A good, if arbitrary, example is the card-levelling system, which attempts to make sure a uniformity of sliver that is not possible by any other means. Other examples would include modern weaving and knitting technologies, as well as highly-controlled dyeing and finishing ranges. One of the most important potential advantages of intelligent automation lies in its ability to provide real-time process monitoring, feedback, and control loops, thereby allowing rapid process correction and minimizing quality faults. Improving the flexibility of a process is a major consideration in deciding whether to automate and, if so, what kind of automation to use.

IBM has made detailed studies of this issue and has been able to relate, at least qualitatively the preferred type of automation to the production volume of one style of item. If production volumes are very low, then human labour is by far the most efficient way to operate a manufacturing process. If production volumes are very low, then human labour is by far the most efficient way to operate a manufacturing process.

If production volumes are very high, then hard automation is the most proper way to operate a process. But for that vast area between very low and very high volumes, intelligent automation or robot is preferred. The practical demonstration of this principle is the huge investment that automobile makers are making in assembly, welding, and painting robots. They must contend with long, but not infinite, runs and there must be model change overs on a periodic basis.

It is said that almost all manufacturing industries are entering an age where they no longer can count on producing the same product day after day, year after year. Flexibility, on a long-term ground at least, is becoming mandatory for any manufacturer of any kind of goods whatsoever. The new technologies that make technological change seem so promising to some and so frightening to others are: computers, robots, and sensing systems. The arrival of the inexpensive computer and its internal essential part, the microprocessor, truly has revolutionized the way in which we perform several daily tasks.

In manufacturing, there is hardly a plant in the country making anything that does not, at the very least, use computers as monitors of the production process and as a source of management information. Much of our current generation of hard automation incorporates small microprocessors for some sorts kinds of intelligent operations, and the availability of low cost but very sophisticated microprocessors has made robots a commercial reality. The idea of the robot is old, but in the past they were too limited in their flexibility to make them very useful. The advent of the low-cost microprocessor permitted robot manufacturers to build, increasing amounts of operational flexibility into them to the point where some robots seem "semi-intelligent".

Current models of robots are able to do several of things. These comprise moving in a number of directions with a number of degrees of freedom, being teachable to do some very complex tasks, being able to react, in a limited way at least, to external events, and having required load carrying capacity and reliability to be suitable for a wide variety of tasks. The technology of robots is developing rapidly and, as it develops, robots become more and more intelligent.

Stress currently is on developing a wide variety of sensory systems comprising vision, tactile sensing, and diverse sort of force and torque sensing; giving them much larger memory so that they can remember and do more complex tasks; providing them mobility so that they no longer have to be bolted to the floor in a particular place to do only one job; teaching them to be self-diagnosing of their own ills; and finally, giving them capabilities for inherent safety so that they can operate in environments with humans without danger to either one.

The chief endeavour in sensing systems at this time is in vision. There exist over 100 manufacturers of vision systems, many of which are aimed at interfacing with robots to provide them the ability to see either two-or three-dimensional objects in black and white' or in various shades or gray or, today, even in colour". Vision can be monocular or binocular and, in the latter case, gives distance judging capability to the robot.

An almost equally rapid area of development is in the taxtile sensing area. Devices exist today that permit a robot to pick up and move very fragile objects such as lamp filaments, using just sufficient force in the gripper to hold the filament without breaking it. Force and torque systems also exist with a general aim of being able to weigh oil object as it is being operated on by the robot.

This has obvious quality control implications in the textile industry, for instance. As the nature of processes in the fiber, textile, and apparel industries is changing so fastly, and because several companies with different philosophies are involved, it seemed appropriate to interview some of the people liable for both the planning and implementation of these changes to determine their views on two fundamental issues.

The first is this: "Regarding current technology, which areas have the most need for improvement and for research?" The second question is equally important: "In about ten years, what new technologies will be important and what research should be done now to aid in their development?" These questions were asked in personal interviews with 27 people in 11 companies, representing the larger and more active companies in the fiber, textile, and apparel industries.

The respondents comprises vice-presidents and directors of manufacturing, vice-presidents and directors of research and development, vice-presidents and directors of engineering, and seven others with diverse titles and job liabilities like purchasing, personnel, product groups, and industrial engineering. These responses connote the probable directions of technological change in the near future for the fiber, textile, and apparel industries.

MAN MADE FIBER INDUSTRY OF THE COUNTRY

With its technological roots in the chemical process industry, the man-made fiber industry in this country has become quite sophisticated in the areas of monitoring, feedback, and control of process parameters. Within the last few years these efforts have been cxtended to give automation, specially in materials handling and warehousing, in the process as an integral part of it.

Although most of these past attempts in automation have been developed internally, and therefore are quite proprietary, there is an increasing tendency to engage outside firms as automation technology becomes more sophisticated. There is a common feeling that, at this time, the technology for materials handling automation after the extruder is quite adequate for present needs and requires only application engineering in most plants to become quite effective. Thus at presently perceived needs for this industry all fall into the general area of monitoring, fecdback, and control, even though these are applied today in comparatively sophisticated forms.

The important philosophical issue is that it is better to prevent defects in as absolute a sense as feasible rather than to manage defects. If a process can be operated with a very high level of defect prevention, then a statistical appraisal of the product is sufficient to make sure that material going to the customer is defect-free. It is believed that very efficient defect prevention needs efficient managing of the polymerization and extrusion steps themselves.

This management takes the form of sophisticated monitoring, feedback, and control of temperatures, pressures, and flow rates in all parts of the process above the spinnerette, and of seeking, ways to monitor and flag potential process defects (like pack pressure build-up or deterioration of spinnerette holes) in such a way that changes can be made before off-quality fiber is produced. The usefulness of threadline monitoring on a continuous ground is not at all agreed upon.

It is argued, for example, that continual monitoring of each thread line is useful when a process is very variable, but in modern fiber extrusion processes, variability is so low that continuous threadline monitoring hardly justifies its cost. Rather, if defects can be reduced to a very low rate, lower than the existing standard of a few defects per thousand pounds of fiber, then statistically determined examination of packages after take-up can catch any process variation that may take place.

Continuous threadline monitoring is used for certain high-value products, i.e., medical hollow-fiber membranes and monofilaments and some industrial monofilaments.

The general aims of the required improvements are to prevent process drift. If process drift can be prevented and if raw material variability is at a suitable minimum, then control of the other process inputs such as temperature, pressure, air flow, flow rates and others, is adequate to guarantee the uniformity of the process. This is a concept that can apply either to filament fiber or staple fiber production. In fact, it is no longer assumed that staple fiber can have a somewhat lower product uniformity than filament.

MODERNISATION AND AUTOMATION OF TEXTILE INDUSTRY

As a whole the textile industry has made major progress process modernization and automation over the last 5 to 10 years. The focus of these programs has been in one of many directions in various companies.Generally new technology has been utilized to reduce the number of process steps, but just as often to reduce the labour content of the process. This is especially evident in the reduction of low skill-high drudgery jobs, but it also applies to the reduction of human intrusion into textile processes, which inevitably is a quality argument.

Examples of these process improvements include the automation of opening rooms, the setting up of chute-feeds and high production cards, the automation, at least in part, of drawing, the introduction of open end spinning, the increasing use of shuttleless looms, the use of automatic systems for handling waste, and the nearly universal use of microprocessor-controlled monitoring and reporting of production variables. Activities in applying existing available technology to textile processes are being pursued in all of the major process categories.

Bale Storage

There is potential for the automatic analysis of incoming bales of fiber, especially if a robot or other automatic device could be involved in the analysis and then direct the bale to the appropriate storage area. Automated bale storage is a labour saving technology that leads to less handling as well as presenting the opportunity of coding each bale for future identification. This is important due to the desirability of tracking the identity of the material in each textile process process from the initial bale through all of the process steps.

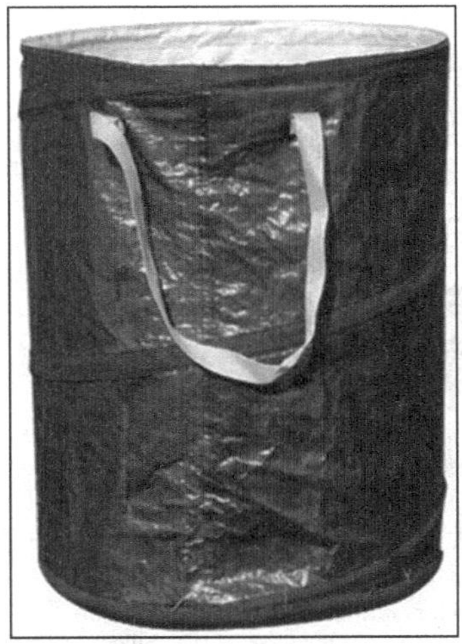

Fig. Bale Storage

This in turn permits one to know the accurate history of all material and all processes and allows the solution of quality control problems on a more rational basis. Cards can be loaded and unloaded automatically and their output can be conveyed automatically to draw frames.

However the requirement to mix the inputs to draw frames in such a way as to insure good blending is potent; these patterns also need to be capable of being changed from time to time.

Fig. Bale Storage Shed

Preparationof Yarn

The issues in yarn formation generally point toward reducing the number of steps wherever feasible. One view is that devising processes that can successfully do this is more potent than devising automated materials handling systems to move materials between currently popular process steps. Generally the ideal yarn manufacture process would take fiber from a bale, convert it to a sliver and move it to a spinning process, for example open end spinning, with automatic transfer of the yarn output either to a warper or to a loom.

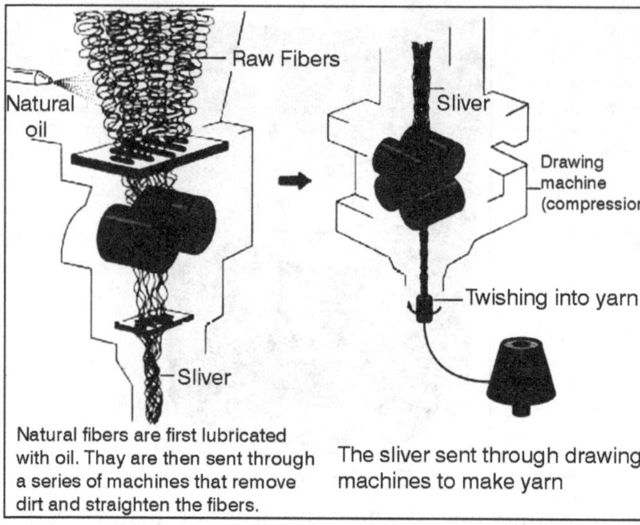

Fig. Preparationof Yarn

Stress also must be placed on computer monitoring both of quality and of production rate, with the quality monitoring being tied to feedback and control mechanisms throughout the many processes. The monitoring technology is already available for drawing and the technology for monitoring either open-end yarn or other types of yarn is on the horizon. Another area of considerable importance with respect to monitoring is to be able to determine the need for machine maintenance by continuous monitoring of yarn from, for instance, open end rotors.

This needs that all ends be monitored regularly and that faulty positions be identified soon. The repair, when it is needed, could be made automatically or, alternatively, that end could be stopped and machine maintenance ordered. A useful system would also monitor and record for long-term gradual deterioration as opposed to short-term problems.

In this whole system, the philosophy is to prevent defective material at its source. In general it is important to measure and to control quality on the spot at every step of the yarn formation process. The computer technology is already available; the real problem seems to be the development of appropriate sensing elements. There are a number of opportunities in present yarn processes for automated materials handling, including the use of robots.

This is especially true for systems that have reduced the number of process steps, i.e., open end or other systems, but it could also be utilised with ring spinning technology in which the automation of roving frames is a current need. Connecting winders to large spinning frames is a good idea, and it was demonstrated at the 1983 ITMA, but winders require to be designed to accommodate some flexibility of yarn count; this perhaps could be accomplished by designing winders with space for extra positions.

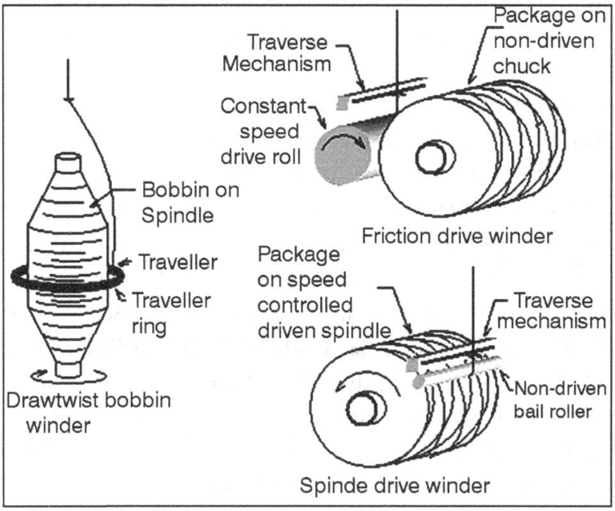

Fig. different stages of yarn processes

This linking technique permits less handling of yarn packages, which is a distinct advantage, and it also allows better package identification and control. Although techniques for the continuous monitoring of different stages of yarn processes are available, knowing where to direct that information or what correclive actions to take still is nat know to a large extent.

Since it is necessary to be able to identify abnormal parts of the process, the real issue is to condition where in a process faults occur, what their relations with other process steps will be, and the feedback loops in which to exercise control.

Preparation of Fabric

The issues in fabric preparation are almost all in the areas of monitoring and control, inspection, and material handling. Modern weaving and knitting machinery allows the choice of several options and a conscious decision of whether to stress productivity or flexibility. The important trend, outside of the design of fabric formation machinery, is the elimination of menial jobs that mostly are related to materials handling.

As an instance, fabric handling with driverless tractors is well-proven now, although this is an area in which potential savings other than labour savings are nat easy to identify; therefore this may remain fairly standard technology for the future.

It also would be desirable to devise systems for automatic doffing of cloth and probably automatic cutting, in conjunction with automatic inspection and grading. Inspection of fabric is thought by several to be a process which should be capable of elimination, if proper process monitoring and control systems can be invented. There is, however, a difference in philosophy depending on whether a greige mill or a dyeing and finishing plant is involved. In the former case it is perhaps more likely that inspection could be eliminated if it were to be carried out later on the finished fabric. Perhaps there will always be a need for some inspection, but it should be automated.

In partical it should be capable of viewing the fabric with both the plant specifications and the customer's specifications in mind, changing the latter as the customer changes and changing the former as the fabric style changes from lot to lot. These issues become more significant as the textile industry as a whole moves to higher quality and output levels as then inspection speeds become the limiting process speeds. Computer-controlled inspection systems also ought to interface with computer-controlled cutting systems to optimize the cutting of first quality yarn lage.

Monitoring in weave rooms specially, but also in other fabric formation areas, is done today almost solely for the objective of generating management information. Although this is an important function of such systems, what is needed, besides, is diagnostic monitoring that will locate and diagnose loom malfunctions, preferably before they lead to producing off-quality material. Such systems would be part of a larger-monitoring system that deals with the flow of raw material into the fabric formation process as well as with fabric formation itself. The development of truly reliable monitoring and control systems is closely association with the development of automated inspection systems, since the latter rely on the assurance that defects are minimized in the process.

DYEING AND FINISHING

Discreate Process

In this area there is a common view that the number of discrete processes in dyeing and finishing needs to be reduced drastically, and as many as feasible should be combined. The goal would be to make dyeing and finishing a truly regular process rather than a series of batch processes, each with its own control and materials handling problems. What this implies is the development of some rather sophisticated monitoring and control systems for dyeing and finishing processes generally. Among these systems, the more potent seem to be the ability to monitor and control the colour of wet fabric, as well as to monitor and control moisture content.

The objective is to be able to predict accurately the colour of the final dry fabric by measuring its characteristics at the moment it is being dyed. Continuous dye ranges now have significant automation but are hampered by their inability to run very small lots efficiently. Systems need to be developed that will permit rapid change over from lot to lot on a continuous range system with a minimum fabric band between the changes. In addition to suitable monitoring and control functions, ways to rapidly alter dye baths in order to change colour fastly must be developed.

Fig. Dyeing Fabrics

The aim is to produce systems with very rapid response times and this will require precision instrumentation for adding chemicals and controlling the parameters of the process. For similar reasons it is extremely important to have absolutely uniform desizing and bleaching.

In this area there my well be applications also for computer-based monitoring, feedback, and control systems. There also are general needs for reducing the energy cost in dyeing by reducing either the setting or the drying needs. And for some products, it would be very useful to have dyeing the last of all finishing steps in order to improve order and warehouse versatility.

Process of Automation in Handling of Materials

From raw materials to the finished product there exist major opportunities for process automation in materials handling. Current applications of robotics are nearly all in this area, for the reason that it seems comparatively easy either to identify labour savings or to accomplish a reduction in menial and sometimes hard to fill jobs related to these processes. A significant concept that accompanies automated material handling is automated identification of the material which, if carried through in all parts of the process permitt the identification of quality control problems regarding their material source.

Much automated materials handling is, or will be, built into the process technology itself; but there remain materials handling problems that will need special technologies, the newest of which are driverless vehicles and robots. On the other hand, it can be considered that automated materials handling is really a second choice to the integration of processes; to the extent that some processes cannot be integrated, materials handling will remain a essential technology.

There is much technology applications development to be accomplished, in the robotics area, which is not yet, in the view of some, suited for most textile applications. Particular issues comprise combining carding with breaker drawing with no intermediate materials handling, the doffing of ring spinning frames, the doffing of cloth at the loom, and the creeling of packages for dyeing, finishing, and warping.

MANUFACTURING PROCESS OF APPAREL

It is easiest to approach this topic by beginning at the front end of an apparel manufacturing process, so the first topic should be computer-aided design. Unfortunately, in its apparel design context, this is a technology that is not quite ready for commercialization. It has the same goal as computer aided fabric design systems, namely, permitting the designer to quickly, and easily test his or her ideas at a computer console with an appropriate viewing mechanism to see the result of those ideas. The problem in apparel design is inevitably that of converting easily and reliably between two-dimensional and three-dimensional information. Technically it would appear easier to take the two-dimensional information from a set of pieces (for instances, for a shirt) and change them to a three-dimensional image of the shirt's final appearance.

However, designers often operate in the reverse mode, that is, they make an illustration of a garment the way they realise it, and then the problem is to convert that picture to a series of two-dimensional pieces to outline the actual manufacturing process. Research progresses and so many theoretical reports are available in this area but, as far as commercialization of this technology goes, progress remains slow.

However, the situation is significantly better in the area of computer-aided marker-making, grading, and cut-order planning. There are many systems on the market that permit an experienced operator to take a standard pattern piece, generate other pieces from it on a graded size basis (i.e., grading) and then to pack all of the necessary pieces on a piece of fabric so as to give maximum fabric utilization. Maximizing fabric use is perhaps one of the greatest driving forces for this technology.

It is estimated that, for plain fabric, fabric wastage can be deducted to about 3 per cent with this technology compared with 5-6 per cent when human beings do the same job unaided. On the other hand, if instead of plain fabrics, one uses patterned fabrics in his process, then fabric wastage can be as high as 25 *per cent* and there is explicit room for improvement here. Although there is no current use of this idea, it appear logical that computer-aided pattern packing ought to be combined with fabric inspection information, performed at the textile manufacturer's plant. Such a combination would allow the apparel maker to cut around any defects that might be present and guarantee that those pieces that were cut were defect free; this would effectively increase fabric utilization. Technologies for cutting fabric are increasingly being debated on a more philosophical basis.

The issues are whether, in fact, the conventional bundle system of fabric cutting and material transport can be retained if other parts of the apparel manufacturing system are to be highly automated. The bundle system uses a spreading machine (which some say has not seen any major improvement in several years) to lay and then cut as several as 100 layers of fabric at once. Thus, this system produces 100 layers of shirt fronts, 100 layers of collars, and 100 layers of pockets and these all are bundled together to give material at each sewing station. The merit of the system is that it is, or at least is thought to be, a very effective and cost efficient way to cut fabric; therefore, it has inherently low costs related to with it.

This assumes that the person who does the cutting, if it is done manually, doesn't make a mistake. One mistake, of course, can ruin 100 shirt fronts. More importantly deducted, from an industrial engineering stand point, is that the bundle system inherently provides a buffer for each of the other steps in the process.

If a sewing machine breaks down or if an operator is sick one day, the manufacturing process can still proceed because of the inherent build up of material available at all stations. The chief demerit of the bundle system is the long in-process time of the item being gathered. It is said that for some articles of clothing the in-plant time is as high as 20-30 days.

This represents an enormous inventory cost and is considered by some to be a major detriment to productivity enhancement. The bundle system also assumes that fabric is identical from one layer to another as the identity of several pieces effectively is lost in the system.

To the extent that fabric is not uniform, then there are inevitable wastages because fabric pieces cannot be properly matched. The alternate system, used in some parts of the apparel industry where fabric is extremely expensive and a cutting mistake cannot be risked, is, for the sake of a better term, the one-layer system.

As the name means, each piece is cut individually in this system and then the problem is to route the pieces in the correct order to their assembly points. The merit of this system, other than the fact that the penalty for cutting errors is minimized, is that this process is actually more amenable to automation and control of the rest of the assembly process.

Besides, such a cutting system permits an enormous reduction of in-process time from 20-30 days to, in some cases, perhaps a few hours. It has obvious advantages. The major disadvantage of the one-layer system is that there is no inherent buffer capability. If a piece is missing or if an assembly station is inoperative, then the whole production process is held up.

On the other hand, it may be feasible to attack this problem by considering the "just in time" manufacturing philosophy that automobile makers use, in which a forward part of the manufacturing process calls to it material just as it is required and no sooner. There is no buffer concept in that philosophy, but it needs adequate material handling systems and reliable parts interchangeability. There are major innovations, some of them now quite old, in the area of cutting fabric.

Computer-controlled cutting, as exemplified by the Gerber cutter, is a technology that has been around for probably 10 years. It is precious machine and is utilized only by large companies that can justify having large cutting departments. In its present form, it takes information directly from the computer-aided marker-maker and cuts pieces from layers of fabric in a very fast and reproducible manner. It also uses a reciprocating knife in most instances, as this has, up to now, been the standard cutting technique.

There are more novel cutting techniques available, each of which has been downgraded in the past because of the demerit they have in working with the bundle system. Laser cutting is a technology that is used to cut out men's suits, for instance.

It can only work really well on one or a few layers of fabric at a time, and it works best on non-thermoplastic fabrics that otherwise would melt at the edges of the cut. (The problem in working with several layers of fabric is that the cut edges tend to fuse together during the cutting process and then they are very difficult to separate.) Water-jet cutting is a technology used in many other fields, such as the shoe industry, the automobile industry, and in a number of industries where complex shapes have to be cut rather rapidly.

Water jets don't work well with the bundle system either, as the power in the jet is severely retarded after the first few layers of fabric. Not with standing, if the one-layer system were to be adopted more uniformly, then either laser cutting or water-jet cutting systems would make considerable sense. Especially this is true as either of these technologies can be combined with robotic guidance of the cutting head. Joining techniques will continue, most people think, to emphasize sewn seams.

Consumers hope certain aesthetic and mechanical attributes of sewn seams, and these are difficult to achieve with any other technology. Modern sewing machines operate mechanically much like the one that How invented, but they have advanced to the point where microprocessor control is an inherent part of several of them.

Microprocessors can control both the stitch pattern and the feeding and ejecting of material once it has been sewn. An emerging technology, not commercial yet, is three-dimensional sewing, in which a sewing head on a robot arm effectively operates in three dimensions on pieces of fabrics that have been placed on a form. There is a considerable amount of research going on in this area. In the sphere of non-sewn techniques, both adhesives and thermal bonding techniques are important. Again, there seem to be applications for robot systems in this area, although the number of seams that can use these-technologies perhaps is limited. As in the textile industry, materials handling in the apparel industry is a major problem, which can be divided into many aspects. The first is materials presentation at the sewing machine itself.

Some apparel companies ascribe as much as 80 per cent of their labour cost to the sewing part of their operations but closer inspection divnlges that actual sewing only contributes perhaps 20 per cent of this cost; the rest is materials handling. There are so many of devices already available to bring material in a more orderly fashion to the sewing machine operator. Two such are made by Gerber and by Eaton. In each case, a rather complex overhead track system is loaded with pieces and partly assembled systems and moved by operator demand to the place where sewing is to take place.

These systems appear to be logical components of more automatic systems for the future. Several have investigated robots or robot-like systems as devices for materials handling, but the chief drawback is the classic limp fabric problem. It seems that it is nearly impossible to design a machine to pick up a piece of fabric, which may have only one of so many weights and stiffnesses, and to operate on it in a reliable and consistent manner. There exist devices which can handle certain classes of fabrics.

The Clupicker is definitely one of these, and there now exist robotic end effectors that are claimed to be able to operate with a wide variety of fabric types. One of these is comprised in the Draper project technology. Nevertheless, until the problems of reliable' materials flow, not only at sewing stations but between sewing stations, is solved, automation in any real sense apparel manufacturing will be severely handicapped. The manufacturing loop requires to be closed with sufficient management information, not only to track a process but also, at least in the ideal case, to determine when a process is reaching trouble.

There are a number of management information packages being sold into the apparel industry now which rely for the most part on operator input. Considering that a sewing machine operator has so many other things to do, one has to wonder just what the quality of the input data to these systems really is. Nevertheless, this represents a significant improvement from the traditional information gathering technology that has been prevalent.

INCREASING USE OF AUTOMATION

The general themes of all of these comments for all three industries are the increasing use of computer technologies, the requirement for the invention of suitable monitoring, feedback, and control loops and the increasing use of automation in materials handling. These trends arise because processes of the future in these industries will most desirably have minimum numbers of intermediate steps and minimum numbers of people involved in these steps. Another general aim is to design processes with less dependence of unit cost on volume, which therefore would permit management more choice of flexibility or mass throughout for any special product at any particular time. In the fiber industry the will be on developing high-speed extrusion processes that can be fully automatic.

Very large packages will be produced of at least 100 pounds, which will mean less inspection. These packages will be loaded directly onto a creel either for warping or directly for weaving or knitting. Boxes and other temporary storage devices will no longer be used. In staple fiber production the fiber instead of being processed into bales, may well be packed directly into a truck and on delivery, transferred to card chute feeds. In the textile industry, there are those who predict common automation of processes, especially materials transfer among carding, drawing, spinning, and fabric formation processes.

Special technology developments may include new spinning technologies that would either replace or refine current open end technology; these will become essential because cotton production is producing increasingly trashier material. Many respondents suggest that replacing the card is a necessary feature of future technologies, largely because cards don't permit for easy material transfer to the next step. An ideal system would send fiber directly to open-end spinning.

If cards are not replaced, provision for loading and unloading them automatically will require to be made and, in traditional systems, sliver cans will be transferred automatically to draw frames. The output from these can be transferred automatically to open end spinning frames and robots may become increasingly useful at this stage.

In fabric production, an ideal situation will permit changing styles on looms much more rapidly than at present, and it will also include the rapid transfer of computer-generated fabric designs to looms for trial runs. Fabric grading will be largely automatic with the ability to distinguish not only sorts of defects, but also their placement within the fabric and to characterise those conditions with reference to either plant specifications or customer specifications.

Such systems will work on any style and any colour of fabric. Dyeing and finishing, but especially dyeing techniques, will be entirely automatic, capable of making nearly instantaneous changes from one shade to another and capable of reproducing shades absolutely. The implications for computer monitoring, feedback, and control of these systems are so many.

Management information will be much more complete and easy to use than it is today. The information collecting used for managerial decision making will be part of the same process monitoring and control information gathering that will be essential to run an error-free process. There also will be a real requirement to be able to estimate cost, labour and material savings, and management time in the process of determining whether a new device or technology is worthwhile at a particular cost.

The emphasis, in general, in the textile industry as well as in the fiber industry will be on upstream quality improvement, which is more likely to produce a higher quality product than automated inspection systems. In the apparel industry, stress will be placed on more efficient materials use, and on speeding up material flow through the factory.

This may be more important in the long term than direct labour savings. However, the cost of inventory is so high that any technology that allowed a garment to pass through the system in a few hours as opposed to several days would be very cost-effective. There will be increasing emphasis on automation in sewing and other joining technologies and material transfer from station to station.

This is the thrust of both of the widely publicized research programs in the automation of apparel manufacturing: the Draper Laboratories programme and the Japanese MITI programme.

Both of these programs use the concept of the "manufacturing cell", in which robots, hard automation, computer-aided design, and computer-driven information flow combine to define a self-contained manufacturing unit. The first is an American Project sponsored by the Tailored Clothing Technology Corporation, abbreviated (TC)2" but loosely called the "Draper" project after the Draper Laboratories where the work actually is being carried out.

The philosophy of this project is that a manufacturing cell exists wherever a sewn seam must be accomplished, and that the significant problems are getting material into the system, performing some orientation and/or parts identification, probably folding the material, sewing it in a reliable fashion, and ejecting the material from the system.

The Draper prototype machine that was completed some time last year (reputed in its present form to cost around $50,000) consists of a number of novel devices to do these things. The important technology appears to many to be in the use of vision systems to identify and orient fabric parts, and in the claimed invention of a robot that can, in fact, reliably pick up single layers of fabric (although in this case it is for folding; i.e., moving one layer over another) and in a sewing head that travels to the place where the seam is to be made.

Each of these operations is microprocessor controlled and, at this time, the process has been designed to do only one thing, yo sew the sleeves of men's coats. The Draper project investigators claim that the technology is adaptable to a wide variety of apparel assembly processes and some of its aspects certainly are, especially the vision system and the materials handling technology.

Some in the apparel industry criticize the' project, partly on the ground that it represents a very costly, very specialized machine, and partly on the basis that nothing has been done to change the sewing technology at all.

A project receiving almost equal publicity in this country is a Japanese project in which a group of Japanese companies under the auspices of the Japanese government (MITI) have banded together to form an organization known as Technology Research Association of Automated Sewing Systems, which has published a very elaborate programme for the entire automation of apparel manufacture—what really amounts to a completely automated factory.

Whereas the Draper project has cost on the order of $5 million so far, the Japanese project is reputed eventually to cost $50 or $60 million, and there are several who believe that it will cost much more than this before it is finished. At this stage, planning and some fundamental work are being done in the areas of sewing preparation, sewing and assembly, fabric handling, and production control.

Thus, all of the apparel manufacturing process is being attacked, with the desired end result being an automated factory composed of so many individual work cells, into each of which material is fed as appropriate and taken out as appropriate, with the entire process operating very much as an integrated system.

While the Draper project emphasis is partly on the reduction of labour content, the Japanese stress is almost completely on the elimination of labour content. Several explain this as the difficulty in Japan of obtaining apparel plant labour, and suggest that this really is an effort to preserve an industry for which a labour force has largely ceased to exist.

The Hereafter

Are we approaching the day when textiles and apparel will be assembled in an automated factory? This is the objective of several manufacturing industries. The traditional definition of the "factory of the future" is one that combines computer-aided design, hard automation, robotics, and computerized information systems to give a self-contained manufacturing cell that has as its input computer generated information and has as its output both computer generated information and a product that is uniformly manufactured to a specific quality standard. Whether or not this all can happen in the textile industries is dependent on other, nontechnical, aspects of these industries.

If any of these levels of automation are to take place, they almost certainly will take place according to company type and, as a result, several predict that the overall textile industry will become divided into two basic sorts of companies, those that are more manufacturing-oriented and those that are more marketing-oriented. The marketing-oriented companies will tend to be design-oriented, specialist companies who place foot market response above production efficiency.

They will react very quickly to the marketplace, they will manufacture high value-added products, and they will use less of the available technology than they might if they were manufacturing larger amounts of their product. However, some technologies will be potent to them, particularly in the area of computer-aided design and in management information systems.

They will tend to judge automation by its abilily to enhance the value that they consider most important, i.e., flexibility regarding market demands. On the other hand, manufacturing-oriented companies will be very large and they will make very large amounts of their product.

They will be prime candidates for large scale automation and perhaps even for the automatic factory. At any rate, they will have to have the ability for great investment of capital and this; as much as anything, will condition whether a company can afford to be manufacturing-oriented or not.

The primary thrusts of these companies will be productivity and quality and, with a view to realise these, they will effort very hard to lower their overall labour costs (not only their direct labour costs) and to lower their in-process time so as to escape the high cost of inventory.

CONTROL OF STOCK AND PRODUCTION

The chief objectives of a manufacturing unit are to provide the right goods, of the right quality, at the right time and at the right cost.

If these objectives are to be attained, it is inevitable that all the productive resources available to the unit shall be operated to the greatest advantage. Materials, machinery. and labour must be used together. in such a way that the rnaximum utilisation of all three is attained. The complexity of modern manufacturing methods is such that it is virtually hot possible to achieve this necessary synchronisation without some systematic approach.

Planning of Production

Planning of production is the systematic organisation of how the productive resources of the unit should be used in the future to achieve maximum productive efficiency. It has been defined as:

The technique of foreseeing or picturing ahead every step in a long series of separate operations, each such operation to be of maximum efficiency, and of so connoting each step that routine arrangements suffice to cause it to happen in the right place and at the right time'.

Function of Production Control

The function of production control is to make sure that the plans that have been made are actually achieved and to report back to management any divergencies in order that remedial action may be taken. The control of production includes all devices, delegations of authority and responsibility, and procedures essential for achievement of production objectives as defined through managerial exercise of the planning functions'.

Objective of Stock Control

The object of stock control is to make sure that all materials, components, and finished goods are available when and where needed. The definition given in the Production Handbook' is as follows: The provision of the needed quantity and quality of material at the required time and place with tne minimum feasible investment'.

PRODUCTION PLANNING AND CONTROL METHOD

There are various ways in which production planning and control can be undertaken. Generally, these methods fall into two categories:

- Some type of numerical analysis.;
- Some form of graphical method.

Numerical analysis can be utilised with good effect when the planning problem is relatively simple. For example, in a spinning mill where the range of counts spun is not very wide, it is feasible to plan the factory output numerically.

In this case, the forward commitments, regarding weight needed on,each count, can readily be converted into equivalent frame-hours and the forward commitments can be expressed in this way. Similarly, in a weaving shed where the range of fabrics is small and the quantities required are large and fairly constant, a numerical analysis of the 'production needs will often prove adequate. However, when the range of products and kinds of machine is more involved, it is usual to adopt some, form of graphical method as a means of production planning and control. In general, the various graphical methods that can be employed all follow the principle of representing time by linear measurement.

In various cases, however, the requirements are such that a planning and control system may well consist of a combination of the two methods of operation, certain information being presented in numerical form and other aspects of the plan being displayed graphically.

One of the favoured methods of machine-scheduling is to use some type of loading board, and this can take many forms. In one of the simpler methods of operation, each machine, or feasibly a group of machines, is represented by a hook on a board, and a ticket is prepared for each job or batch to be produced. The tickets are loaded on the hook representing the machine on which the material will be processed.

As the work is executed, the tickets are removed from the board, so that at any time the forward load for each machine can be seen by examination of the tickets. In other cases, a loading board may be a comparatively complicated piece of apparatus, the complexity of the manufacturing operation being essentially reflected in the planning that is required. There are many proprietary planning and control systems available; these range from card-index systems with colour-code flashes that cannote the progress achieved to various forms of wall charts that represent the plan in some form of-flexible vertical or horizontal scale.

It is not feasible in a chapter of this nature to enumerate all the methods available for production planning and control, and it must be emphasized that it is rarely feasible to take some system that has been developed, for another plant and introduce it as it stands into one's own factory. In the introduction of a planning system into a manufacturing unit, the inevitable approach is to establish explicitly exactly what information is required, and then to devise some method of presenting this in the simplest possible way, so that the system can work efficiently and effectively, but at the same time as economically as possible.

There is always the danger, which must be constantly borne in mind, that, instead of the planning system's being a service to the management, the factory can become a slave to the planning system. Anyone who has the for production planning and control should therefore make himself completely familiar with all the different methods that are available in order that he is in a position to select the particular approach that 'will best meet the needs of his own firm.

Stock Control Methods

Stock control is realised by straightforward ledger accounting, in which goods received and goods issued are recorded, and the balance of stock in hand is always available. These records may be kept numerically, in which case they will take the form of some type of stock account, which will be maintained in a ledger, or on some form of card system.

Due to its flexibility and successfully used in different factories. Such illustrations are available in other publications, however, and in any case it is the writer's belief, as has already been said, that there is no system of production planning and control or of stock control that has universal applicability. Although it is possible to follow certain broad common principles, every plant has such special problems of its own that an effective system must be individually tailored to meet the exact requirements of the factory into which it is introduced.

It is proposed to consider two special problems and suggest feasible methods of approach which, it is hoped, will illustrate the application of the principles that have been outlined. It is not suggested that the methods that have been adopted are the only ways in which control could have been instituted, and they are intended only as an indication of the way in which the problem may be approached.

PRODUCTION PLANNING AND CONTROL'S INSTANCE

A manufacturing unit may be weaving a wide range of fabrics to individual customers' exact needs in a shed comprising several kinds of loom. It may be necessary for both warp and weft yams to be specially dyed to the shades vanted for each order. Let us assume that the management wishes to introduce a production-planning and control system covering the weaving shed that will as far as feasible keep it fully informed as to the position regarding current and future production. A list of the items of information that the system will be expected to provide may run as follows:

- The actual production that has been obtained from the department compared with that which was planned.
- The delivery dates that may be quoted against any enquiries for new business on any certain type of loom.
- The position of each individual customer's order.
- The forward-loading position for each kind of loom.
- The effect on other orders if it becomes necessary to change the planned production for any certain lot.
- The current rate of production against any certain customer's order.
- The warp-yarn delivery position.
- The number of pieces that have been delivered to the customer.
- The planned date of entry into the loom of each warp making up an order.
- The weft-yam position.
- The number of pieces that have been woven against any order to date.
- The number of pieces held in stock ready for delivery.

The outstanding balance of any order still to be delivered. Let us now follow through one method of approaching such a planning and control problem. First of all, we must setup the information that will be needed and the sources from which it will be obtained.

The technical department tells us that the warp yarn will be bought in hank, dyed to pattern, cone-wound, section-warped, and hand-loomed. The weft yam will be bought on cone, dyed on cone and pirn-wound. There will be ten 100-yd. pieces on each beam, and the cloth will be woven in loom type C. We have agreed with the purchasing department that we must permit the spinner three weeks from receipt of order for delivery to dyer, and that a further three weeks will be needed for the dyeing of the yarn.

The work-study engineer informs us that three weeks will be needed for warp preparation from hank, and that the yam should be on beam two weeks before it is required for the loom to allow for feasible congestion in the looming department. The standard weaving-production rate for this particular quality is 200 yd./week. As the manufacture of this order progresses, the production-control section will feed back the following information:

- The date on which each beam is prepared.
- The date on which the warp yarn is despatched to the dyer.
- The date on which the weft yarn is obtained from the dyer.

- The date on which the warp yarn is received from the dyer.
- The date on which the weft yarn is despatched to the dyer.
- The dates on which pieces are completed on each loom.
- Details of deliveries made to the customer.

If this list is examined, it can be seen that items A.F. are mainly related to the planning of the productive capacity, whereas items G.N. refer to thee progress of the customer's order. It is unlikely that all the information that is needed could be given on one schedule without its becoming so complex that interpretation would be too difficult to be Practical.

Since the information falls naturally into two categories, let us therefore decide to use two planning media, the first covering the weaving-shed activity and the second -recording the progress of the customer's order. Of the different methods available, some form of loading board based on Gantt-chart principles will undoubtedly be most suitable for planning the loom production. Management has expressed a demand for warp and weft control. This is effected by a combination of the vertical datum lines and the small blocks on each card.

The datum lines consist of cords stretching down the board, and these are moved forward with the passage of time, so that the datum line represents the existing date and the other control cords maintain their relative position.

We have been informed that it is necessary to have the warp yam on beam two weeks before it is needed for the loom and the 'warp-on-beanl' line is therefore two weeks ahead of the current date. The yarn must be received from the dyer three weeks before it is due on beam to allow for warp preparation, and the line marked 'warp from dyer' is therefore a further three weeks ahead. The 'warp-to-dyer' line is fixed in the same way.

The distance between the current-date line and the continuity line represents the total time needed for yarn delivery, dyeing, and warp preparation. Against the warp entry on the card, there are three small blocks, and below that are two blocks covering weft. When the warp yarn is delivered to the dyer, the first of the small blocks marked warp' is filled in; when the warp yam is received from the dyer, the second of these blocks is marked; and, when the warp yarn is run onto the weaver's beam, the third is covered.

When the weft yarn is delivered to the dyer, the first of the weft blocks is marked; and, when the weft yarn is received from the dyer, the second block is fined in. For ease of interpretation, it is advisable to use various colours for each of these blocks and to match them to the colours of the cords representing the various datum lines. As pieces are woven, the appropriate spaces on the top edge of the cards are blocked in, so that it is possible at any time to see how several pieces have been woven from each beam. If we now refer back to the information that management needs, we find that it can be obtained as follows:

- The forward-loading position for each type of loom will be shown, and any break in continuity can be clearly seen. For example, if there were no further orders for loom type C beyond the one shown on the loading board, there would be serious loss of continuity on all looms separate from loom 146, and even here immediate action is needed if we are not to wait for orders.
- Delivery dates for new business can be quoted to commence from the date when the beam cards end on the loading board, provided that they extend beyond the continuity line. Due adjustment must be made of course, for production running behind the planned schedule.
- By comparing the filling in of the warp squares with the different datum lines, the warp-yam delivery position can be seen at a glance. All beams that are planned to go into work between the date line and the 'warp-on-beam' line should have all three squares filled in. If we look at loom 176, we can see that the' second beam for this loom is because of to start weaving in less than two weeks, but the third square has not yet been filled in, which connotes that this beam has not been prepared.
- By rearranging the beam cards, it is feasible to assess the effect of any change in the planned production.
- By comparing the marking of pieces woven on the top edge of the card with the existing date line, it is immediately apparent if looms are running to schedule. We can see that loom 176, for example, is running almost properly to plan, whereas loom 299 is over a week behind.

Like, all beams planned to go into the loom before the 'warp-from-dyer' datum line line should have the second square' filled in.

Here we can see that the yarn for the third beam on looms 86 and 94 is overdue from the dyer. The weft-yarn position is shown by the two' blocks opposite 'weft' on the card. In the instance, both these blocks have been fined in to show that the yarn has been despatched to, and received from, the dyer. When the beam cards have been placed in their appropriate positions on the loading board, some details of the plan are transferred to the customer's order sheet. The loom numbers into which beams are planned to be woven are inserted, together with the date on which these looms are due to start weaving. The number of pieces on each beam is indicated and the actual date on which they go into work is entered.

As pieces are woven, the card on the loading board is marked, as mentioned above, and the date is inserted in the appropriate square under the 'pieces woven' heading on the order sheet. In the lower portion of the sheet can be seen the deliveries that have been made to the customer, together with the balance that has yet to be delivered. The information that management has asked for regarding individual customers' orders may now be obtained as follows:

- The date on which weaving actually commences on each warp is recorded.
- The over-all position of each certain order is represented on the order sheet.
- All the 54 pieces have been delivered.
- The planned date of entry into the 100m of each beam is shown.
- The outstanding balance yet to be delivered is 146 pieces, or 14 622 yd.
- It can be seen from this sheet eight looms are weaving on this order and at a standard production of two pieces per loom per week, the existing rate production is 16 pieces per week.
- No pieces are therefore at present held in stock awaiting delivery.

The approach followed in the establishment or this control system may be summarized as follows:

- Effort to find a method of presentation that will present management with all the detail it requires as simply and clearly as possible, with the minimurn use of time and effort.
- Reach agreement with management as to the precise information it requires.
- Establish properly what facts will be needed, and from which source they will be obtained.

Stock Control Instance

It may be that a firm manufactures goods against stock orders and sales are made from inventory. At any time, under these situations, management may wish to know the precise detail of any stock order as follows:

- Number of items that have been sold and are not yet manufactured;
- The balance of a manufacturing order that is yet unsold;
- The number of items in stock that have been manufactured but not yet delivered to the Customer.
- The balance of the manufacturing order that is still to be made
- Variety of items that have been sold and have not yet been delivered to the customer; and
- This information can be received by a numerical method.

The method of control, however, is somewhat cumbersome and comprises a great deal of calculation, especially when, as in this case, goods are counted in dozens and twelfths, as is customery in such trades as shirts., towels and certain kinds of bedding, etc. If one casts around for a simpler method of presentation that will comprise no numerical work the feasibility of employing some form of graphical representation springs to mind.

6

Labor Process in Textile Industry

Ever since individual capitalists began employing large numbers of workers simultaneously, there have been attempts to increase their productivity, or output of labor. Because capital purchases the labor power of workers rather than their product of labor, that labor power is under the direct control of the buyer.

Because capitalists purchase means of production as well as labor power to produce and sell commodities, they seek to organize production in such a way that surplus value, their source of profits, is maximized. Changes in the labor process have been made by both the introduction of detail work into the division of labor (manufacture) and the continuous introduction of new technology (machinofacture).

Many writers, such as Braverman, Edwards, Marglin, and the Brighton Labor-Process Group, identify capitalist-initiated changes in the labor process as rooted in the need for capitalists to cut costs of production and maintain management control over labor.

Capitalists need control over the labor process because of the specific nature of the capitalist mode of production: the commodification of labor so that entrepreneurs purchase the labor power, not labor of workers. Various conditions, a socially useful product, may interfere with the capitalist goal of maximizing profits. By dividing the labor force into details as well as introducing machinery that incorporates the skills of workers, management is in a position to replace more highly skilled workers with lesser skilled workers and thus save labor costs, although a small number of new, highly skilled jobs can be by-products of this process. These innovations of the labor process make workers more replaceable and subject to manage-ment control, thereby enhancing capital's bargaining position over labor.

Other writers, such as Magaline, Gartman, and Geschwender and Levine, stress that increasing mechanization and rationalization of the labor process is intrinsically linked to the struggle between capital and labor. The antagonism derived from the relationship between capital and labor provides another impetus for changes in the labor process. As such, this antagonism is the motor force or prime mover behind innovations in the labor process. Rising labor costs in an increasing competitive environment exacerbate this antagonism.

Despite the differences in theoretical emphasis concerning changes in the labor process, most writers agree on their effects. Mechanization and routinization increasingly tend to deskill most workers and/or eliminate some workers from production, and the relative bargaining position of labor, having been made more replaceable, is weakened visà-vis capital. This chapter examines the specific strategies that textile entrepreneurs adopted during the post-World War II period to both enhance their relative position vis-à-vis textile labor and increase their competitiveness in the market.

Textile production was one of the first industries to be industrialized. Indeed, the Industrial Revolution centered around this industry. Although the transformations that had occurred in textiles were earlier than in other industries, they stagnated in the early 1900s as other industries became more mechanized and capital intensive. Davis et al., writing on U.S. manufacturing from 1910 to 1960, note little technological change in the textile industry during this period. However, the postwar period is characterized by substantial change and innovations.

Despite changes that have occurred, textile production remains relatively labor intensive when contrasted with other industries. The amount of capital invested per production worker is low when all sectors of industry are examined. Because textile production is labor intensive, labor costs comprise a large part of the price of textile commodities.

Of the cost of these commodities, 57 per cent is related to production, the remainder consisting of distribution and marketing costs. Of this 57 per cent, over 36 per cent has been attributed to direct labor costs, and concern with reducing this latter percentage has prompted textile entrepreneurs to mechanize the labor process.

TEXTILE PRODUCTION

The post- World War II textile industry experienced increasing mechanization of the labor process to meet the challenges of international competition and rising labor costs. This was one of several possible ways to meet these challenges. For example, the apparel industry, a closely related labor industry, reestablished "sweatshops" in the 1970s in urban centers.

The conjuncture of a booming economy, revision of depreciation rates, trade agreements limiting imports, and beneficial tax adjustments encouraged textile firms to invest in new equipment during the 1960s. Plant and equipment expenditures rose from $330 million in 1961 to as much as $820 million in 1966. Toward the late 1960s and continuing into the 1970s, however, textile machinery prices rose approximately 25 per cent. Coupled with high inflation rates in other areas (utilities, cotton, etc.), capital investment, especially in spinning and weaving, slowed, although outlays in knitting and carpet mills increased. In addition, mechanization in the 1970s lagged behind the 1960s because higher costs fostered a decline in the profit margins of the industry. Approximately two to three cents per dollar was the typical rate of profit on investment in the early 1970s, compared with the four- to five-cent margin in the 1960s.

The cyclical business patterns and the continued long-term debt of the industry kept most of the industry's firms from continuing reinvestment in new capital.

Four specific factors facilitated the adoption of new technology in the textile labor process. First, increasing tendencies toward concentration and centralization of textile capital (through increasing mergers and acquisitions) gave large mills greater amounts of capital to invest. Second, the post-World War II economic boom created conditions that temporarily improved the financial capabilities of textile firms, especially the largest mills. Third, the research and development support provided by state agencies as well as research and development gains in the chemical industry (i.e., synthetics) were translated into technological changes in the way textiles were produced. Fourth, state policies in the early 1960s permitted faster tax write-offs on capital investment, thereby encouraging capital investments.

At the same time, increased mechanization occurred unevenly. The industry still experienced a relatively high degree of vertical fragmentation. Because of the uneven development within the textile industry, larger mills were the ones in a position to adopt new forms of technology. Thus technological changes in the labor process, with their concomitant effects on textile workers, were for the most part confined to the larger mills. Larger mills became the pacesetters in the industry, and thus are the focal point of this analysis.

The period for investment in new plants and technology that led to transformations in the textile industry generally occurred between 1962 and 1968. In the 1970s, investment in new technology was largely confined to machinery rather than to the construction of new plants. This was due in large part to the industry's compliance with the Occupational Safety and Health Administration (OSHA) cotton dust standards.

GENERAL TECHNOLOGICAL DEVELOPMENTS

Many technological and organizational innovations introduced in the textile industry during the 1960s affected textile workers. These include the development of synthetic fibers, the introduction of electronic instrumentation and computers, the new design in mill architecture, and adoption of new technology and new hiring practices within the industry. Synthetic fibers were first manufactured on a large scale during this period. They had always been potentially attractive commodities because of their superior wear, easy care, wash and dry features, and so on, but their limitations were that they did not have the warmth or comfort of natural fibers. However, starting in the 1960s, synthetic fibers were produced in a manner similar to natural fibers, permitting synthetics to obtain many of the properties of spun yarn. A treatment of synthetics called "texturizing" put a "crimp" into the fiber through a heat process, which provided bulk to it.

There are two types of synthetics: cellulosic and noncellulosic. Cellulosic fibers are comprised of inert substances, in the form of carbohydrates, which act like cell walls. Rayon, acetate, and triacetate are cellulosic fibers (i.e., made from wood pulp). Noncellulosic fibers (petroleum derivative) include nylon, acrylic, polyester, and polypropylene. The blending of synthetics with natural fibers has become commonplace.

The production of synthetics is oligopolistic the three largest producers account for over 60 per cent of the industry's total sales. Its oligopolistic character is due, in part, to high capital requirements, the high level of technological knowledge and experience necessary, the necessary commitment to extensive research and development expenditures, and the enormous marketing effort required to sell these new fibers. All these requirements obstruct the entry of new firms in the market.

The introduction of electronic instrumentation and computers into the textile labor process is further important innovations, although their applications are unevenly diffused within the industry. These developments in various stages of production are integral parts of the industry's transformation to a more capital-intensive labor process. Large and small mills have incorporated electronic controls into newer machinery, but computer-controlled systems are generally found only in larger mills. Computers have been used on a fairly wide scale in the dyeing and finishing aspect of the labor process, since they permit the greatest combination of flexibility for consumers and economy for manufacturers in determining the cheapest combination of dyestuffs to match colour.

Once programmed, the system will work unattended, producing the sixteen best recipes and costings required to match a customer's dye, so that the most economical mix can later be chosen. This is no mean feat when it is realized that for one colour using only three shades, it is often possible to obtain twohundred thirty-eight recipes alone.

Computers have also been introduced in accounting, inventory control, and production planning by the larger mills. Programming is used in the mills for materials handling, management reporting, and process control. Operated from a control room in a plant with the aid of prepunched cards, supervising tasks such as inspection can be mechanized. Moreover, the computer can both predict and compute the weight of racks and discern the appropriate speed of the machinery carrying them for such a weight. Scanners, run by computers, detect defects. An increasing number of mills have computer-controlled operations. One mill has a computer-directed handling system that moves material from truck unloading to the placement of heavy beams of yarn in racks. This innovation eliminates many of the unskilled workers in the mill. Robots have been introduced in a smaller number of mills. For example, in one mill, the robot on command places the filling yarn on a conveyor belt for delivery to the weaving shed.

Computers have also been employed to run looms at different locations, even across continents. The automated machine can only make sense when it becomes part of an automated mill where input of yarn scheduling, flow of materials, knitting, processing and shipping are all computer controlled and coordinated for optimal efficiency. In a setup like this, the knitting machine will no longer be an entity of its own but form a unit plugged into one integral manufacturing system.

Such a setup would affect supervisors and forepersons, as they would be dispatched by computers with clear instructions on the nature of a given problem. The introduction of computers thus grants mill owners and managers greater control over the labor process as well as reduces the need for machine operatives, maintenance personnel, and unskilled laborers. Skill requirements for repair technicians and electricians are upgraded.

The architecture of textile mills built since the 1960s differs from earlier architecture, reflecting technological changes. Newer mills are longer, one-storied plants. Unlike older multistory plants that required moving of partially completed materials to other stages of the production process on other floors, one-storied mills facilitate combining and integrating of various processes.

Mechanization of these processes can reduce the amount of time needed for moving materials from one stage of production to another, thereby increasing productivity and potentially lowering labor costs. Hence the onestoried mill permits certain types of mechanization in the labor process, such as that of moving and loading materials that could not readily have been introduced in multistoried mills. What were once formally segregated processes of production can now become an integrated unit.

The Whitin Machine Works introduced the first systematized mills, in which each part of textile production was designed to mesh with others, keeping manual handling of materials at various stages to a minimum. This process reduced total labor cost per pound from 9.0 cents per dollar to 3.81 cents.

Leslie Fay, a New York-North Carolina firm, integrated this production of yarn to knitwear itself, thus linking the production of textiles with apparel. Although the larger firms are in a better position to afford new plant expenditures on the one-story model, this type of facility is clearly in vogue. New capital to the industry can thus be highly mechanized from the start.

The introduction of new, more technologically advanced machines has also changed the labor process and displaced workers. Changes in machinery can be classified into three general types.

The first type merely modified existing machinery to affect speed, cleaning, and so forth. This form of change is the least expensive to adopt and the most widely diffused throughout the textile industry. It has led to faster and larger production capacity machines. Changes of this kind reduce the number of machines and machine operatives required for the same output. The second type of technological change includes new machinery and changes that are connected to already existing machinery.

This kind of change tends to be concentrated in larger, more modern mills, and includes the mechanization of materials handling through conveyor belts, chutes, chute-fed carding, and builtin maintenance, which reduces the need for some maintenance workers.

Finally, the third type entails radical innovations, which integrate distinct processes. Examples of these include open-end spinning, the shuttleless loom, and automatic doffing. This type of change is almost exclusively confined to the largest firms. Most technological changes in the post-World War II period are confined to improving existing machines; only the very largest companies invest in path-breaking machinery.

Most technolgical changes that U.S. textile firms have adopted originated from the so-called revived foreign textile industries, particularly Japan and West Germany although some have originated in socialist countries such as Czechoslovakia.

For example, Japanese textile machinery manufacturers have instituted new spinning units that require only 5 machine attendants, compared to the 40 attendants characteristic of machines produced by U.S. textile machinery manufacturers. An administrator for the Saco-Lowell Research and Development Centre in Clemson, South Carolina, advised local firms to look abroad for certain types of new machinery that would drastically reduce the number of laborers required for an equal output of commodities. U.S. textile machine manufacturers lack the financial resources and the productive capacities to keep pace with foreign competition.

As firms have become increasingly concentrated and centralized, the need for greater standardization among various mills as well as maximization of output has led to the hiring of professionally trained managers. This shift of emphasis in hiring practices has changed the nature of management-labor relations. Prior to 1945, and even for the early years of the post-World War II period, managers were largely recruited from either the family of the mill owner or someone who had moved up the occupational ladder within the mill. In such an arrangement, especially in the mill communities, there were personal ties between managers and workers; workers many times grew up with their supervisors.

Worker loyalty to the mill owner was assured by these "paternalist" relationships, and efforts were made to keep workers and their families employed in the mill even when this hiring was not the soundest decision using profitability criteria. The new managers stressed impersonal relationships with workers and used profitability criteria (e.g., Taylorism, cost-benefit analysis) in their decisions concerning the employment status of workers. Thus the shift in managerial strategy corresponded with the transformation in the labor process.

IMPACT OF CHANGES IN TEXTILE LABOR PROCESS

Textile production reduces raw cotton into thread or yarn, which is woven or knitted into cloth, and then finally dyed and finished. These are the major steps of cotton yarn and fabric textile production. This series of operations can occur separately in specialized mills or be integrated within a large mill. There is a relatively high degree of similarity between these processes and those of synthetic yarn and fabric manufacture.

Opening and Blending

Smaller mills rely on the traditional method of taking several bales of cotton and having a worker lift the cotton layer by layer into the opening machines. These machines loosen and clean the fiber as well as blend it to assure proper mixing. This type of work, which employs less than 5 per cent of workers in the older mills, is unskilled and involves heavy lifting and handling of materials. Newer mills have been mechanized to eliminate manual operations.

Besides mechanization of the handling of new fiber, opening and blending machines are larger and faster, and the quality of the yarn is increased.

Picking

The older, traditional mills have mechanized conveyor belts that move cotton from the opening and blending machines to picking machines that remove impurities as well as parallelize the fibers, forming them into sheets. These sheets are then rolled into large rolls called "laps." The operator of the picker machine ("picker tender") removes ("doffs") the laps from the machine, weighs them, and manually transfers these heavy rolls onto a hand truck.

A trucker then takes the laps to the next operation. The modernized mills have mechanized all of these processes. The "pick" lap rolls are moved along conveyor belts, weighed and possibly rejected mechanically, and moved to the next process. Thus the work of the unskilled trucker is eliminated altogether, and the once heavily demanding work of the picker tender is reduced to starting and stopping the conveyor belts and monitoring the belts for problems.

Carding

After picking, the laps of fiber go to the carding machines, which straighten and further parallelize the fiber via a surface of bent wires into thin strands called "card slivers." These card slivers are coiled into cans and then manually trucked to the subsequent process.

In the old mills, the "card tender" is a semiskilled machine operative who moves along many carding machines, feeding the laps into them, removing the coiled sliver, repairing broken ends of fiber, and cleaning the lint and waste that the cards took off the fiber. Changes in the carding room during the 1960s increased output fourfold from the previous decade. There is greater instrumentation (e.g., stop-motion devices), and automatic suction devices on the machine absorb much of the lint and waste that the cards pick up.

A card tender thus spends less time on this task, only 10 to 15 per cent of total work time, rather than the 65 per cent spent in traditional mills. The need for maintenance (i.e., card fixers and card grinders) is substantially reduced by automatic lubrication. Because the card tender spends less time cleaning the cards, he or she tends three to four times as many cards as in the traditional mills.

More mills are adopting a continuous opening-blending-carding operation into the labor process. This operation known as "direct-feed" or "chute-feed" carding eliminates handling of the fiber from one machine to another. Direct-feed carding eliminates this entire picking process. This operation no longer requires picker operations and other unskilled and semiskilled labor. In addition to productivity increases, OSHA cotton dust level requirements are met in this process.

Drawing

Drawing frames continue to process the fibers by combing several fiber strands and increasing parallelism and uniformity. In the older mills, the drawing frame tender, who is considered a semiskilled worker, looks after many rows of drawing frames, repairs broken ends of sliver and other malfunctions, replaces the cans into which the strands are coiled, and cleans the machines. Whereas four to five drawing frame tenders are required in the older mills, newer mills require only one worker for the same output. The introduction of automatic cleaning and conveying devices, as well as larger and faster machines, requires considerably fewer machines and workers.

Roving

The roving process further reduces the fibers into smaller strands, inserts a slight twist into the fiber, and then winds the fiber onto a bobbin. Because bobbins are manually doffed and trucked to the next process in the older mills, men have predominated as machine operatives in this process. During the 1960s automatic cleaning devices and automatic doffing were introduced. These developments accompanied a feminization of this job category. In some larger mills, the roving process has been eliminated altogether.

Spinning

This process draws out the fibers and twists the strands into yarn. They are then wound into smaller bobbins, which are manually doffed. Workers in this process comprise approximately 20 per cent of all mill machine workers. In older, smaller mills, the spinner's job consists of "manually loading the machines with roving bobbins, twisting the ends of woven yarn from the spent bobbin, repairing breaks, cleaning the machines, and patrolling the long lines of machines carefully watching for broken strands or nearly exhausted fibers."

Whereas the spinners are largely female workers, doffing of full bobbins is for the most part performed by male workers. Newer mills have eliminated doffers altogether, as in the earlier processes where automatic doffing has been introduced. There are automatic cleaning devices as well, so that the machine operative "simply starts the doffing machine and patrols the rows of spindles to check for malfunction."

Automatic doffers, first produced in Japan, have been used by only a few of the larger firms to integrate this process with opening and blending and carding, thereby eliminating the picker tender altogether. These very expensive machines blow the opened and blended fiber through pneumatic ducts directly through the carding machines. While output is not greatly increased, sharply reduced labor costs lead to increased labor productivity.

A new technique, introduced in Czechoslovakia, and only recently used in U.S. mills, is "rotor" or "open-end spinning." Open-end spinning increases productivity fourto fivefold by eliminating the conventional spindle and integrating the roving, spinning, and winding processes. Space, maintenance, and cleaning requirements are reduced, so that fewer unskilled and semiskilled workers are needed.

Winding and Warping

This process prepares the yarn for use on the loom for the weaving or for the knitting machines by transferring it from the bobbins into larger packages. Whereas the loading and doffing of the materials as well as some knotting are manually operated in the older mills, these processes have been mechanized in the newer mills. The newer machines are larger and faster and, for the same output, reduce the number of workers required by 60 per cent.

Weaving

Weaving interlaces threads on the loom to form a fabric, called a "grey (sometimes greige) cloth." Roughly one-half of all machine operators in textile mills that perform all stages of textile production are in the weaving room. Machines in older mills are smaller and slower than those found in the newer and larger mills; a weaver in a more modern mill may control 200 looms compared with 150 to 160 for the older mill. Machine maintenance has been reduced through automatic cleaning and lubrication devices. The shuttleless loom has been a major innovation in the process. This loom operates without a shuttle, bobbin, or bobbin feeding mechanism; requires only 30 per cent of the labor used on the conventional loom; and increases productivity by 200-300 per cent. Because fewer preparatory processes are required for weaving with the use of the shuttleless loom, further costs are reduced. Moreover, unhealthy noise levels have been reduced by this innovation. The shuttleless loom is very expensive and has become increasingly diffused within the industry. Another innovation in weaving, which originated in Japan, is the water-jet loom. It, too, raises productivity; however, it has been adopted less frequently in the United States since it can only weave continuous filament synthetic fibers.

Finishing

This process singes, desizes, bleaches, dyes, prints, preshrinks, and so forth, the fabrics. In older mills, these techniques were discontinuous, taking lengthy periods of time. Newer operations introduced in the mid-1960s combined many of these formerly disjuncted processes into a continuous system. An open bleaching system, found only in a few mills, takes only seven minutes compared with nine hours in the older mills. This process has been aided by developments in the chemical industry. Computers have been introduced in these processes as well.

Another innovation of the postwar textile industry, which is outside the spinning and weaving processes, is the tufting of carpets. Tufted carpet has almost completely replaced loom-woven carpet manufacture; whereas in 1950 there was practically no tufted carpet production, by 1965 tufted carpets and rugs accounted for approximately 87 per cent of all soft floor covering mill shipments. By 1980, tufted carpet accounted for 96 per cent of the total quantity of carpets and rugs produced.

EFFECTS OF MECHANISATION ON TEXTILE WORKERS

Mechanization in the textile industry has had two basic effects on textile workers. First, many workers have been displaced from the labor process altogether. These were mostly unskilled workers engaged in carrying the unfinished goods from one machine to another, as well as machine operatives. Second, textile workers have become increasingly deskilled and routinized, as more and more skills have been incorporated into the machine.

Although the U.S. textile industry has been a machine industry rather than a craft industry since the nineteenth century, gradations of skill levels remained. At the onset of the post-World War II period, the skilled workers were the most militant and willing to join a union, and increasing deskilling in the labor process made their skills more dispensable. Hence increased mechanization in textile production displaces textile workers as well as requires a less skilled work force.

Much of the technological change in the production of textiles has been aimed at reducing or eliminating what textile management considers superfluous labor.

That is, many of the manual operations performed by textile workers, whether creeling (loading), doing (removing), repairing or fixing broken fibers, operating machine controls, or cleaning the machines have been either transferred in varying degrees to the machine, or a new machine has been built with equivalent functions. These technological changes have also eliminated certain processes, such as the trucking of semi finished goods.

Moreover, a given amount of textile products can be made with a fewer number of machine operatives. As a result, many textile workers have been displaced from the labor process. Between 1950 and 1976, textile employment was reduced from 1,245,200 to 919,000, a decline of 27 per cent, while output increased by 124 per cent. An undetermined number of textile workers who lost their jobs during this period were displaced as a result of imports of textiles and apparel.

Increased mechanization of the labor process has also further routinized textile workers. There is less work for the machine operative to perform that requires manual dexterity and accumulated knowledge concerning the operation of a machine.

Thus the content of many workers' jobs has been altered. As many of the machines utilized in the newer mills incorporate the manual operations of the textile worker, the textile machine operator patrols the machines more of the time. As fewer workers are required, machine operatives tend to patrol a larger number of machines as well.

A comparison of the distribution of a spinner's work over three decades reveals that he or she spends more time as a machine tender. In 1950, over four-fifths of a spinner's time consisted of no patrolling operations; by 1970 this was reduced to roughly one-half. At the same time, the number of machines tended more than doubled. These changes can be attributed to the introduction of automatic loading, winding, and cleaning into production.

Weavers were at one time the most highly skilled and prestigious craftsmen in the industry. Even after the automatic loom eliminated many of their traditional skills, their jobs retained considerable variety and interest.

The same is true for spinning. Before the realignment of jobs, a spinner blew off her frame with an air hose, cleaned off "picked" tolls, set in new bobbins of roving in her machines, put on ends after doffing and put up broken ends of yarn. At present, she performs only the latter of these five functions. The rest are consigned to other specialized workers or are taken care of by automatic processes.

On the whole, less physical labor and manual dexterity are required of textile workers. However, the machine operative tends a greater number of machines, which operate at a faster pace. The importance of detecting malfunctions or errors increases as the costs to the firm increase, so that pressure on the worker intensifies and anxiety mounts.

Barkin writes that as textile workers spend more of their time tending machines, at the same time they acquire more responsibility because they now have to be capable of following written and oral instruction. However, the Bureau of Labor Statistics observed that for machine operatives, the training time is not longer and in many cases is actually less than before.

Increased mechanization has had two beneficial effects on textile workers. First, there is regularly maintained air conditioning. Second, the amount of cotton dust in the air is reduced through automatic cleaning, although the problem still exists.

Textile workers have sought government support to decrease their exposure to cotton dust, the cause of byssinosis, commonly referred to as "brown lung". While A. Uris, a textile industry consultant, envisages the decline of the production worker, he also foresees a rise in designers, repair people, and laboratory technicians. The process of routinization of labor has contradictory tendencies; it also creates certain skilled jobs necessary in keeping the complex machines running. As machine speed, capacity and automaticity increase fewer semi-skilled operatives, such as spinners and weavers, are required for the same production. Similarly, unit requirements for laborers are being sharply reduced in all modernized mills as materials handling is more mechanized or two operations are combined.

In contrast, skilled workers such as mechanics and fixers are becoming more essential as textile machines increase in complexity and continuity. New technologies, such as computer processing and pollution control devices, may also require more technically trained personnel than the textile industry has generally required. Among white collar workers, every major group — professional and technical, managers and officials, clerical and sales — has been increasing as a proportion of total employment.

Hence, while certain unskilled forms of work such as loading, doffing, and trucking continue to be eliminated via mechanization, and machine operatives, who are also slowly being displaced from the labor process, continue to spend more of their time merely patrolling their machines, certain mechanical jobs increase in skill, as more expertise is required to keep more complex machines running. The highest wages are paid to those workers who do not work the machines directly, but rather set up or repair them.

Furthermore, the importance of technicians and engineers has increased, particularly in the larger mills where computerization has appeared. In the larger mills, as many as seven technicians are required just to keep the air conditioners functioning. Finally, the larger and faster machines have given textile firms greater flexibility in meeting market demand, and this flexibility, coupled with increasing concentration and centralization of capital within the industry, has increased the relative proportion of white-collar workers. Accountants, salespeople, stenographers, secretaries and typists, and other clerical workers have become high-growth nonproductive occupations within the industry.

PRODUCTIVITY

The transformation of the labor process through mechanization not only displaces and further routinizes textile labor, it also increases productivity. Increments in productivity increase the competitiveness of firms in the markets, which is particularly important for U.S. firms that must compete with foreign mills located in low-wage,

Third World areas. Despite uneven development within the U.S. textile industry, available data indicate that the introduction of new forms of mechanization in textile production has translated into huge productivity gains. Moreover, the textile industry had the highest growth rate of labor productivity between 1948 and 1979 among the 20 major manufacturing industries in the United States.

In addition to increasing diffusion of technology, another factor contributing to increased productivity is the greater concentration within the industry. There is uneven development within the textile industry. Large modernized firms coexist with smaller conventional ones, although the larger mills continue to assume a larger share of total production each census year.

During the 1960s, 1970s, and 1980s, domestic and foreign competition, coupled with downturns in business cycles, led to closings of many smaller, less-efficient mills. As a result, production has increased while the number of firms has been reduced. Hence productivity rates, as affected by the closing of smaller firms, would tend to rise, for these smaller mills probably brought the rates down for the industry as a whole in earlier years.

Despite the sparse data, the technological changes that transformed the labor process have indeed increased productivity and have made U.S. firms competitive when compared to those of Europe. However, data to show how U.S. firms have improved their relative standing in competition with European firms since the early 1960s are lacking, as are any comparisons with Asian textile industries.

7

Methods used in Treatment of Textile Wastewater

TEXTILE INDUSTRIES WASTEWATER

Textile industry is one of the most water and chemical intensive industries worldwide due to the fact that 200-400 liters of water are needed to produce 1 kg of textile fabric in textile factories. The water used in this industry is almost entirely discharged as waste. Moreover, the loss of dye in the effluents of textile industry can reach up to 75%.

The effluents are considered very complex since they contain salt, surfactants, ionic metals and their metal complexes, toxic organic chemicals, biocides and toxic anions. Azo dyes are regarded as the largest class of synthetic. Approximately, 50–70% of the available dyes for commercial applications are azo dyes followed by the anthraquinone group.

Azo dyes are classified according to the presence of azo bonds (–N=N–) in the molecule *i.e.*, monoazo, diazo, triazo etc. and also sub-classified according to the structure and method of applications such as acid, basic, direct, disperse, azoic and pigments (Bhutani, 2008). Some azo dyes and their dye precursors are well-known of high toxicity and suspected to be human carcinogens as they form toxic aromatic amines. Different physical, chemical and biological as well as the various combinations of pretreatment and post-treatment techniques have been developed over the last two decades for industrial wastewaters treatment in order to meet the ever-increasing requirements of human beings for water. Though there are numerous studies published in this field, most of the techniques adopted by these researchers are uneconomical, ineffective or impractical uses. Recent studies have demonstrated that heterogeneous photocatalysis is the most efficient technique in the degradation of colored chemicals. These studies used titanium dioxide and/ or zinc oxide in the photolysis processes.

The large bang gap of titanium dioxide and zinc oxide (~ 3.2 eV) put a limitation of using these semiconductors in photocatalytic degradation under natural weathering conditions. Only a small part of the overall solar intensity could be useful in such photodegradation processes. However, the existence of dye on the surface of catalyst reduces the energy required for excitation and then increases the efficiency of the excitation process by extending its absorption in the visible region of the spectrum.

There are several factors to choose the appropriate textile wastewater treatment method such as, economic efficiency, treatment efficiency, type of dye, concentration of dye and environmental fate. There is no general method for the treatment of textile industrial wastewater. Wastewaters from textile industry contain various pollutants resulting from various stages of production, such as, fibres preparation, yarn, thread, webbing, dyeing and finishing. Mainly three methods are used for the treatment of textile industrial wastewater.

These are:

1. Physico-chemical methods.
2. Advanced oxidation methods.
3. Biological sludge methods.

The main operational methods used for the treatment of textile industrial water involve physical and chemical processes. However, these techniques have many disadvantages. These disadvantages include Sludge generation, high cost, formation of bi – products, releasing of toxic molecules, requiring a lot of dissolved oxygen, limitation of activity for specific dyes and requiring of long time. In recent years, advanced oxidation processes (AOPs) have gained more attraction as a powerful technique in photocatalytic degradation of textile industrial wastewater since they are able to deal with the problem of dye destruction in aqueous systems.

PHOTOCATALYSIS AND PHOTOSENSITIZATION

Photocatalysis is defined as the acceleration of a photoreaction in the presence of a catalyst, while photolysis is defined as a chemical reaction in which a chemical compound is broken down by photons. In catalyzed photolysis, light is absorbed by an adsorbed substrate. Photocatalysis on semiconducting oxides relies on the absorption of photons with energy equal to or greater than the band gap of the oxide, so that electrons are promoted from the valence band to the conduction band:

$$\text{Semiconductor} + hv \rightarrow h^+ + e^-$$

If the photoholes and photoelectrons produced by this process migrate to the surface, they may interact with adsorbed species in the elementary steps, which collectively constitute photocatalysis. The numerous gas- and liquid-phase reactions photocatalysed by TiO_2 and ZnO have been reviewed. There is a general agreement that adsorption is necessary since the surface species act as traps for both; photogenerated holes and electrons, which otherwise recombine. In addition to participating in conventional surface reaction steps, adsorbed dye molecules assist in the separation of photoholes and photoelectrons, which may otherwise recombine within the semiconductor particles. A major factor affecting the efficiency of photocatalysis process is electron/hole recombination. If the electrons and holes are used in a reaction, a steady state will be reached when the removal of electrons and holes equals the rate of generation by illumination. Recombination and trapping processes are the de-excitation processes which are responsible for the creation of the steady state, if no reaction occurs.

There are three important mechanisms of recombination:

1. Direct recombination.
2. Recombination at recombination centers.
3. Surface recombination.

There are different types of semiconductors, whose band gaps range between 1.4 and 3.9 eV, *i.e.*, it could be excited with a light of 318–886 nm wavelengths. This means that most of the known semiconductors could be excited by using visible light. However, not all these semiconductors could be used in the photocatalytic reactions. Bahnemann *et al.*, (1994) report that the most appropriate photocatalysts should be stable towards chemicals and illumination and devoid of any toxic constituents, especially for those used in environmental studies. The authors also explaine that TiO_2 and ZnO are the most commonly used in photocatalytic reactions due to their efficient absorbtion of long wavelength radiation as well as their stability towards chemicals. Other semiconductors like WO_3, CdS, GaP, CdSe and GaAs absorb a wide range of the solar spectrum and can form chemically activated surface-bound intermediates, but unfortunately these photocatalysts are degraded during the repeated catalytic cycles involved in heterogeneous photocatalysis.

The illumination of suspended semiconductor in an aqueous solution of dye with unfiltered light (polychromatic light) leads to the possibility of the existence of two pathways:

1. In the first pathway, the part of light with energy equal to or more than the band gap of the illuminated semiconductor will cause a promotion of an electron to conduction band of the semiconductor and as a result, a positive hole will be created in the valence band. The formed photoholes and photoelectrons can move to the surface of the semiconductor in the presence of light energy. The positive hole will react with adsorbed water molecules on the surface of semiconductor producing $^{\bullet}OH$ radicals and the electron will react with adsorbed oxygen on the surface. Moreover, they can react with deliquescent oxygen and water in suspended liquid and produce perhydroxyl radicals ($HO_2^{\bullet}$) with high chemical activity (Zhao and Zhang, 2008). The processes in this pathway can be summarized by the following equations:

$$h^+ + OH^- \rightarrow \dot{O}H$$

$$h^+ + H_2O \rightarrow H^+ + \dot{O}H$$

$$e^- + O_2 \rightarrow O_2^{\bullet -}$$

$$O_2^{\bullet -} + H^+ \rightarrow HO_2^{\bullet}$$

$$HO_2^{\bullet} + O_2^{\bullet -} \xrightarrow{H^+} H_2O_2 + O_2$$

$$H_2O_2 \rightarrow 2\dot{O}H$$

$$\text{Dye} + \text{Semiconductor}\ (h^+_{VB}) \rightarrow \text{Dye}\cdot^+ + \text{Semiconductor}$$

$$\text{Dye}\cdot^+ + O\cdot_2^- \rightarrow DO_2 \rightarrow \text{degradation products}$$

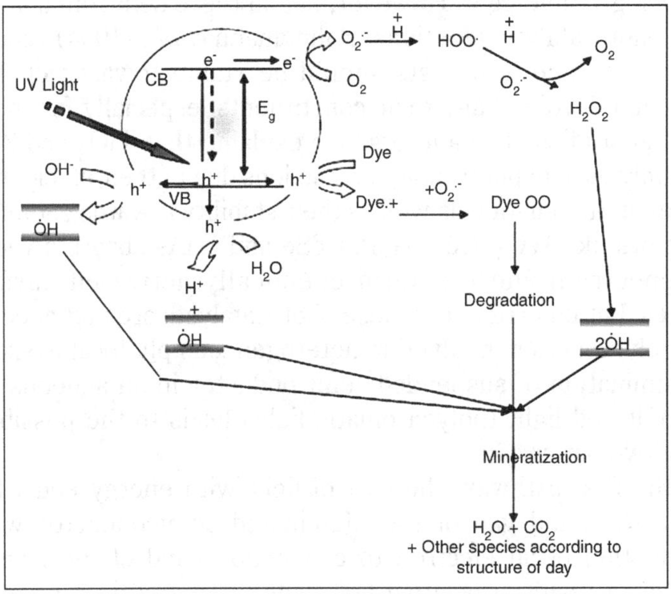

Fig. Dye/Semiconductor/UV Light System

2. In the second pathway, the other part of light with energy which is less than the band gap of the illuminated semiconductor will be absorbed by the adsorbed dye molecules. Dye molecules will be decolorized by a photosensitization process. The photocatalytic decolorization of dyes, which is described as photosensitization processes, is also characterized by a free radical mechanism. In this process, the adsorbed dyes molecules on the surface of the semiconductor can absorb a radiation in the visible range in addition to the radiation with a short wavelengths. The excited colored dye

(dye*) (in the singlet or triplet state) will inject an electron to the conduction band of the semiconductor (Hussein and Alkhateeb, 2007). The processes in this pathway can be summarized by the following equations:

$$\text{Dye} + \text{hv}(\text{VIS or UV reagion}) \rightarrow {}^1\text{Dye}^* \text{or}\ {}^3\text{Dye}^*$$

$${}^1\text{Dye}^* \text{or}\ {}^3\text{Dye}^* + \text{semiconducto} \rightarrow \text{Dye}^{\cdot+} + e^- (\text{to the conduction band of semiconductor})$$

$$e^- + O_2 \rightarrow O_2^{\bullet-} + \text{Semiconductor}$$

$$\text{Dye}^{\cdot+} + O\cdot_2^- \rightarrow \text{Dye } O_2 \rightarrow \textit{degradation} \text{ products}$$

$$\text{Dye}^{\cdot+} + HO\cdot_2 \text{ (orHO)} \rightarrow \textit{degradation} \text{ products}$$

$$\text{Dye} + 2\dot{O}H \rightarrow H_2O + \text{oxidation products}$$

$$\text{Dye}^{\cdot+} + \bar{O}H \rightarrow \text{Dye} + \dot{O}H$$

$$\text{Dye}^{\cdot+} + H_2O \rightarrow \text{Dye} + OH\dot{+}H^+$$

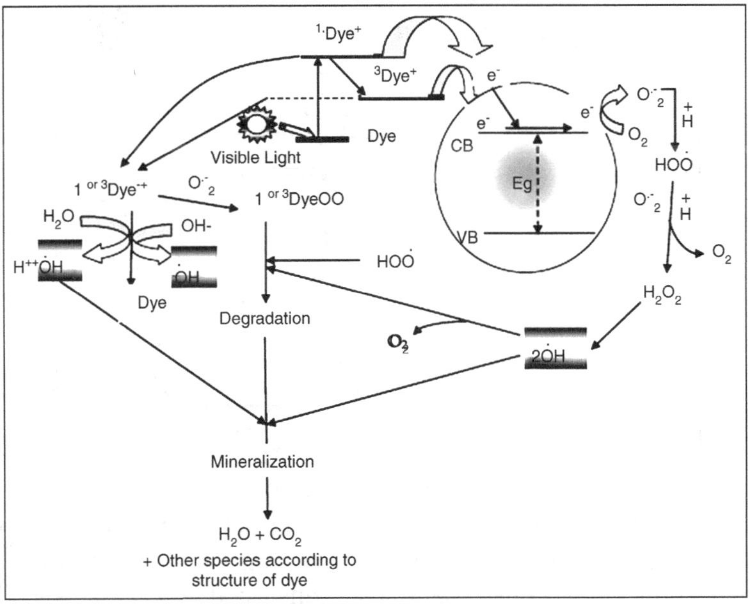

Fig. Nonregenerative Dye/Semiconductor/Visible Light System

The mechanism above is favoured by nonregenerative organic dye where dye/semiconductor/visible light system and the sensitizer itself degrade. However, in regenerative semiconductor system, the following mechanism may be followed:

$$\text{Sen} + \text{TiO}_2 \leftrightarrow \text{Sen} - \text{TiO}_2$$

$$\text{Sen} - \text{TiO}_2 + \text{RX} \leftrightarrow \text{Sen} - \text{TiO}_2......\text{RX}_{(ads)}$$

$$\text{Sen} - \text{TiO}_2 + \text{O}_2 \leftrightarrow \text{Sen} - \text{TiO}_2......\text{O}_{2(ads)}$$

$$\text{Sen} - \text{TiO}_2 \underset{hv}{\overset{hv\ \text{visible light}}{\rightleftarrows}} \text{Sen}^* - \text{TiO}_2$$

$$\text{Sen}^* - \text{TiO}_2......\text{O}_{2(ads)} \rightarrow \text{Sen}^+ - \text{TiO}_2 + \text{O}^-_2$$

$$\text{Sen}^+ - \text{TiO}_2 + \text{O}_2^- \rightarrow \text{Sen} - \text{TiO}_2 + \text{O}_2$$

$$\text{Sen}^* - \text{TiO}_2\ \text{Sen}^+ - \text{TiO}_2\ (e^-_{C.B})$$

$$\text{Sen}^+ - \text{TiO}_2\ (e^-_{C.B}) \rightarrow \text{Sen} - \text{TiO}_2$$

$$\text{Sen}^+ - \text{TiO}_2\left(e^-_{C.B}\right)......\text{RX} \rightarrow \text{Sen}^+ - \text{TiO}_2 + \dot{\text{R}}\text{X} + \text{X}^-$$

$$\text{Sen}^+ - \text{TiO}_2\left(e^-_{C.B}\right)......\text{O}_{2(ads)} \rightarrow \text{Sen}^+ - \text{TiO}_2 + \text{O}_2^-$$

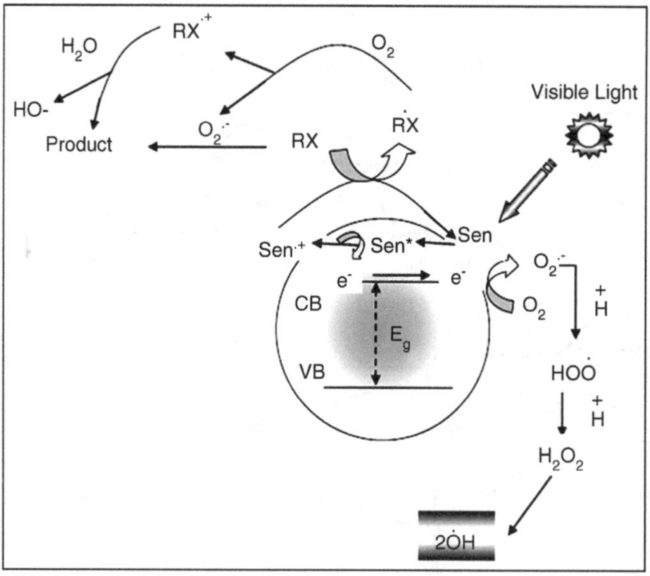

Fig. Regenerative Dye/Semiconductor/Visible Light System

Cho *et al.*, (2001) conclude that there is no direct electron transfer between an excited sensitizer and CCl_4 molecules, in homogeneous solution, and the existence of semiconductor is essential for sensitized photocatalysis. Platinum supported on titanium dioxide acts as an excellent sensitizer and could have practical advantages as a mild and convenient photocatalyst for selective oxidation processes. The addition of Rhodamine-B, as sensitizer to TiO_2 dispersion system increases the rate of photooxidation properties. The authors explained that due to the fact that more light absorbed by Rhodamine-B between

460-580 nm, then the energy transfer from sensitizer to TiO_2 or to any other active species and hence promote the photocatalytic activity of titanium dioxide.

ADVANCED OXIDATION PROCESSES

Glaze *et al.*, (1987) define Advanced Oxidation Processes (AOPs) for water treatments as the processes that occur near ambient temperature and pressure which involve the generation of highly reactive radicals,especially hydroxyl radicals (•OH), in sufficient quantity for water purification. Advanced oxidation processes can also be easily defined as techniques of destruction of organic pollutants from wastewaters. These processes include chemical oxidation processes using hydrogen peroxide, ozone, combined ozone and hydrogen peroxide, hypochlorite, Fenton's reagent, ultra-violet enhanced oxidation such as UV/O_3, UV/ H_2O_2, UV/air, wet air oxidation and catalytic wet air. Hydroxyl radicals are strong reactive species, which are capable of destroying a wide range of organic pollutants. Table shows hydroxyl radical as the second strongest oxidant.

Table. Standard Reduction Potential of Common Oxidants against Standard Hydrogen Electrode

Oxidant	E° (V)
Fluorine (F_2)	3.03
Hydroxyl radical (•OH)	2.80
Atomic oxygen (O)	2.42
Ozone (O_3)	2.07
Hydrogen peroxide(H_2O_2)	1.78
Hydroperoxyl radical ($O_2H\cdot$)	1.70
Potassium permanganate ($KMnO_4$)	1.67
Hypobromous acid (HBrO)	1.59
Chlorine dioxide (ClO_2)	1.50
Hypochlorous acid (HClO)	1.49
Hypochloric acid	1.45
Chlorine (Cl_2)	1.36
Bromine (Br_2)	1.09
Iodine (I_2)	0.54

The attack of organic pollutants by hydroxyl radicals occurs via the following mechanisms:

- *Electron transfer from organic pollutants to hydroxyl radicals*:

$$\dot{O}H + RX \rightarrow RX^{\cdot +} + OH^{-}$$

- *Hydrogen atom abstraction from the C–H, N–H or O–H bonds of organic pollutants*:

$$\dot{O}H + RH \rightarrow R^{\cdot} + H_2O$$

- *Addition of hydroxyl radical to one atom of a multiple atom compound*:

$$\dot{O}H + Ph \rightarrow HOPh^{\cdot}$$

FUNDAMENTAL PARAMETERS IN PHOTOCATALYSIS

In semiconductor photocatalysis of industrial wastewater treatment, there are different parameters affecting the efficiency of treatment. These parameters include mass of catalyst, dye concentration, pH, light intensity, addition of oxidizing agent, temperature and type of photocatalyst. Other factors, such as, ionic components in water, solvent types, mode of catalyst application and calcinations temperature can also play an important role on the photocatalytic degradation of organic compounds in water environment.

EFFECT OF TYPE OF CATALYST

Haque and Muneer (2007) observe that Degussa P25 is more reactive for degradation of a textile dye derivative, bromothymol blue, in aqueous suspensions than other commercially available photocatalysts types of titanium dioxide, namely Hombikat UV100, PC500 and TTP. They explain the high activity of Degussa P25 is due to composing of small nano-crystallites of rutile dispersed within the anatase matrix.

The band gap of rutile is less than that of anatase and as a result electron will transfer from the rutile conduction band to electron traps in anatase and the recombination of electrons and holes will be reduced. Hussein, (2002) reported that anatase has higher photoactivity than rutile due to the difference in surface area. Decolorization percentage of real textile industrial wastewater on rutile, anatase, and zinc oxide shows that the activity of different catalysts falls in the following sequence:

$$ZnO > TiO_2 \text{ (Anatase)} > TiO_2 \text{ (Rutile)}$$

ZnO is more active than TiO_2 due to the absorption of wider spectrum light (Sakthivela *et al.*, 2003). However, the amount of zinc oxide required to reach the optimum activity is two times more than that for titanium dioxide (anatase or rutile) (Hussein and Abass, 2010 a).

In another study, Hussein *et al.*, (2008) observed that ZnO is less active than anatase when the same weight of catalysts is used for photocatalytic degradation of textile wastewater.

Akyol *et al.*, (2004) reported that ZnO is more active than TiO_2 for the decolorization efficiency of aqueous solution of a commercial textile dye due to the band gap energy, the charge carrier density, and the crystal structure. Decolorization efficiency of real textile industrial wastewater in the presence and absence of catalyst and/or solar radiation was also investigated. The results indicate that the activity of different catalysts fall in the sequence:

$$ZnO > TiO_2 \text{ (Anatase)} > TiO_2 \text{ (Rutile)} > \text{in the absence of catalyst} = \text{in the absence of solar radiation or artificial radiation} = 0$$

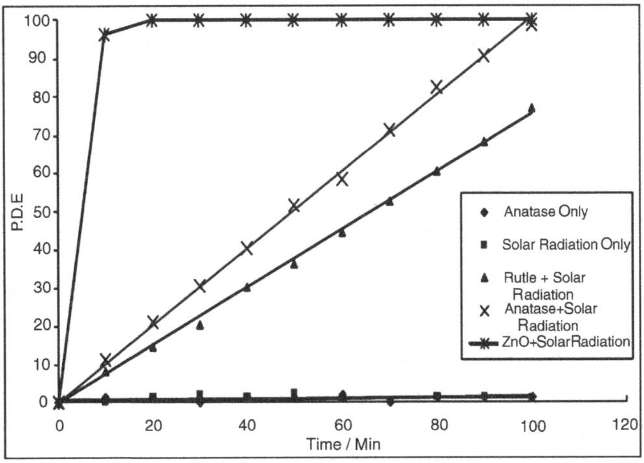

Fig. Photocatalytic Decolorization of Real Textile Industrial Wastewater at Different Conditions.

These results also indicate that there has been no dark reaction. Incubations of colored industrial wastewater without solar radiation and/or without catalyst has been performed to demonstrate that decolorization of the dye is dependent on the presence of both; light and catalyst.

EFFECT OF MASS OF CATALYST

Photocatalysts dosage added to the reaction vessel is a major parameter affecting the photocatalytic degradation efficiency. Photocatalytic degradation efficiency increases with an increase in catalysts mass. This behaviour may be due to an increase in the amount of active site on surface of photocatalyst particles. As a result, an increasing the number of dye molecule adsorbed on the surface of photocatalyst lead to an increase in the density of particles in the area of illumination. The extrapolation of Hird's data (Hird, 1976) indicates that only 7.5 mg of TiO_2 was sufficient to absorb all incident 366 nm radiation. It follows that the mass effect must be caused by changes in the effective utilization of the absorbed radiation rather than by the increased absorption. Photocatalyst with small particles are more efficient than larger particles.

This behaviour may be due to (Hussein, 1984):

- Photoholes and photoelectrons generated in the bulk would have fewer traps and recombination centers to overcome before reaching the surface.
- A greater proportion of material would be within the space charge arising from depletive oxygen chemisorptions, which favour exciton dissociation and photohole migration to the surface.

Hence, increasing the catalyst's mass will increase the concentration of the efficient small particles within the illuminated region of the reaction vessel. The direct proportionality between photocatalytic degradation efficiency and catalyst loading is real within low concentrations of photocatalyst where there are excess active sites reaching plateau reign. The plateau is reached when this effect can no longer increase the overall efficiency of utilizing incident radiation.

The plateau region is achieved, the activity of photocatalytic decolorization decrease with increase of catalyst concentration for all types of catalysts. This behaviour is more likely to emanate from variation in the intensity of radiation entering the reaction vessel and the way the catalyst utilizes that radiation. Light scattering by catalyst particles at higher concentration lead to decrease in the passage of irradiation through the sample leading to poor light utilization. Deactivation of activated photocatalyst molecules colliding ground state molecules with increasing the load of photocatalyst may be also cause reduction in photocatalyst activity.

Photocatalytic decolorization efficiency (PDE) % of real textile industrial wastewater has been investigated by employing different masses of TiO_2 (anatase or rutile) or ZnO under natural weathering conditions for 20 minutes of irradiation.

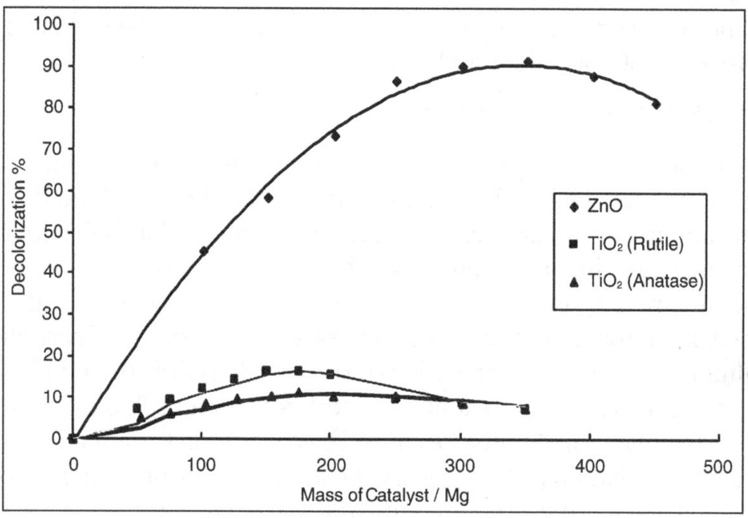

Fig. Effect of Mass on Photocatalytic Decolorization Efficiency of Real Textile Industrial Wastewater

The results in all cases indicate that the decolorization efficiency increases with increase in catalysts mass and then it becomes constant. It is clear from consideration of the catalyst concentrations at which the activity plateau were achieved that the mass effect does not depend upon the type of dye and source of irradiation.

Moreover, plateau regions were achieved and then the activity of decolorization decreased with increasing catalyst concentration, for all types of catalysts used in this project.

EFFECT OF PH

Aqueous solution pH is an important variable in the evaluation of aqueous phase mediated photocatalytic decolorization reactions. pH change effects the adsorption quantity of organic pollutants and the ways of adsorption on the surface of photocatalyst (coordination). As a result, the photocatalytic degradation efficiency will greatly be influenced by pH changes. Zero Point Charge (pHzpc), is a concept relating to adsorption phenomenon and defined as the pH at which the surface of an oxide is uncharged. If positive and negative charges are both present in equal amounts, then this is the isoelectric point (iep). However, the zpc is the same as iep when there is no adsorption of other ions than the potential determining H^+/OH^- at the surface.

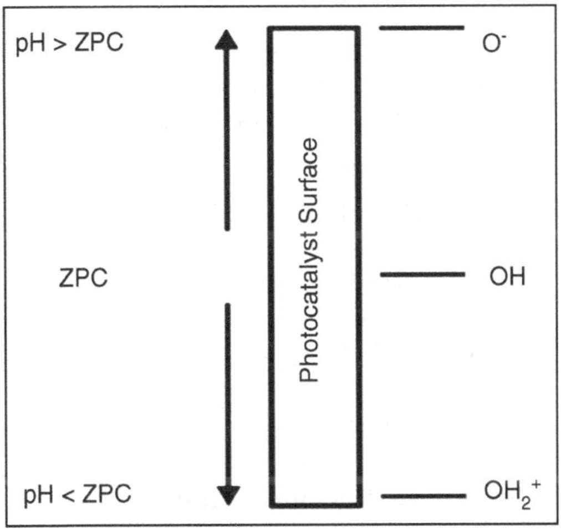

Fig. Effect of pH on ZPC.

In aqueous solution, at pH higher than pHzpc, the oxide surface is negatively charged and then the adsorption of cations is favoured and as a consequence, oxidation of cationic electron donors and acceptors are favoured. At pH lower than pHzpc, the adsorbent surface is positively charged and then the adsorption of anions is favoured and as a consequence, the acidic water donates more protons than hydroxide groups. Infrared spectroscopy study of Szczepantiewicz *et al.*, (2000) shows that the TiOH sites are the major electron traps when TiO_2 is illuminated. The distribution of other species ($TiOH_2^+$ and TiO^-) with changing pH has been proposed by Kormann and co-workers. Figures show that, at pH below ZPC the surface is mostly positively charged and TiOH sites increase as pH increases and reach maximum value at ZPC of semiconductor. However, $TiOH_2^+$ as pH increases and reaches zero value at ZPC. At pH higher than ZPC the density of TiO- groups on the surface start to form and reached 100% value at pH 14. The importance of pH during the reaction is not less than that of initial state. The formation of intermediate products, sometimes, changes the pH of aqueous solution and as a result, it affects the rate of photodegeradation (Galvez, 2003).

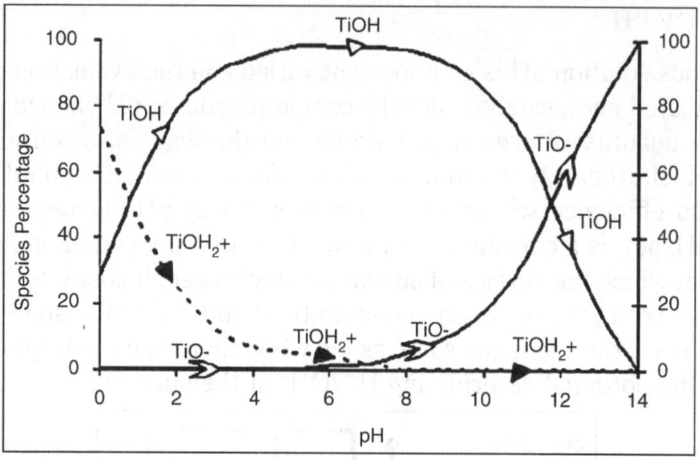

Fig. The Distribution of TiO_2 Surface Species at Different pH

At pH above and below pHzpc, the surface of zinc oxide and titanium dioxide are negatively or positively charged according to the following equations:

$$ZnOH + H^+ \leftrightarrow ZnOH_2^+$$

$$ZnOH + OH^- \leftrightarrow ZnO^- + H_2O$$

$$TiOH + H^+ \leftrightarrow TiOH_2^+$$

$$TiOH + OH^- \leftrightarrow TiO^- + H_2O$$

Effect of pH on Photocatalytic Decolorization of Bismarck Brown R

Under the determined experimental condition with initial dye concentration equal to 10^{-4} M, ZnO dosage 3.75 $gm.L^{-1}$, light intensity equal to 2.93 $mW.cm^{-2}$ and temperature equal to 298.15 K, the effect of change in solution pH on decolorization percentage has been studied in the range 2-12. The decolorization per cent has been found to be strongly dependent on pH of solution because the reaction takes place on the surface of semiconductor. The decolorization percentage of Bismarck brown R increases with the increase of pH, exhibiting maximum decolorization at pH 9.

Under the determined experimental condition with initial dye concentration equal to 10^{-4}M, TiO_2 dosage 1.75 $gm.L^{-1}$, light intensity equal to 2.93 $mW.cm^{-2}$ and temperature equal to 298.15 K, the effect of change in solution pH on decolorization percentage has been studied in the range 2-10. TiO_2 (DegussaP25), TiO_2 (HombikatUV100), TiO_2 (MillenniumPC105) and TiO_2 (Koronose2073). It was observed that the decolorization percentage strongly depends on the pH of solution because the reaction takes place on the surface of semiconductor. The decolorization percentage of Bismarck brown R increases with the increase of pH, exhibiting maximum decolorization at pH that is equal to 6.61, 6.54, 6.75, 6.63 for TiO2 (DegussaP25), TiO_2 (HombikatUV100), TiO_2 (MillenniumPC105) and TiO_2 (Koronose2073), respectively.

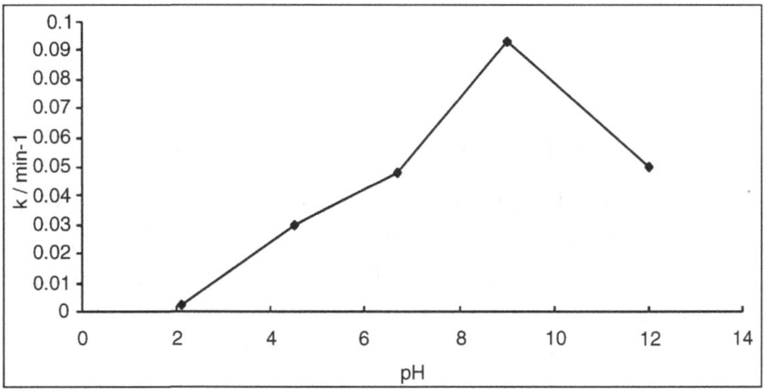

Fig. Effect of pH on Photocatalytic Decolorization Efficiency of Bismarck R Brown on ZnO

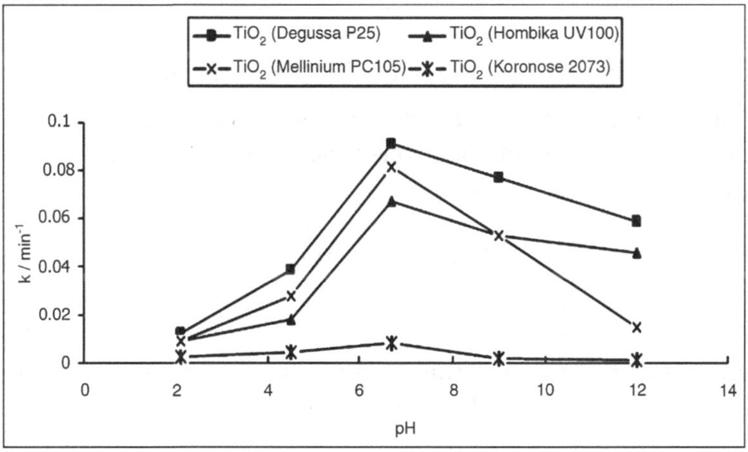

Fig. Effect of pH on Photocatalytic Decolorization Efficiency of Bismarck R Brown on Different Types of TiO_2

This behaviour could be explained on the basis of zero point charge (ZPC). The zero point charge is equal to 9.00 for ZnO and 6.25 for TiO_2 (Degussa P25). With the increase of the pH of solution, the surface of catalyst will be negatively charged by adsorbed hydroxyl ions. The presence of large quantities of adsorbed OH^- ions on the surface of catalyst favour the formation of •OH radical. However, if pH is lower than ZPC, the hydroxyl ions adsorbed on the surface will be decreased and, therefore, hydrogen ions adsorbed on the surface will increase and the surface will become positive charged. Both the acidic and basic media leave an inverse impact on the photodecolorization efficiency because of the decrease of the formation of the hydroxyl radical. The decolorization of Bismarck brown R decreases dramatically at strong acid media (pH = 2.1) for ZnO.

This could be explained due to photocorrosion of ZnO as shown in the following equations:

$$ZnO \xrightarrow{hv} e^{-}_{CB} + h^{+}_{VB}$$

$$ZnO + 2h^{+}_{VB} \rightarrow Zn^{2+} + \frac{1}{2}O_2$$

EFFECT OF LIGHT INTENSITY

Egerton and King, (1979) show that square root of light intensity depends on the activity of titanium dioxide for different wavelengths of light. However, this relationship cannot be applied to all range of light intensities. The primary electronic processes which occur in the absorption of photons with energy equal or greater than the band gap of semiconductor are:

$$\text{Semiconductor} + hv \xrightarrow{k_1} (h-e)\text{exciton}$$

$$(h-e) \xrightarrow{k_2} h^{+} + e^{-}$$

$$h^{+} + e^{-} \xrightarrow{k_3} \text{Radiationless Re combination}$$

For photocatalysis processes, it is necessary that the excitons dissociate and the photoholes and photoelectrons reach the catalyst surface where they are trapped by surface species:

$$h^{+}OH^{-}_{(S)} \xrightarrow{k_4} \dot{O}H_{(S)}$$

$$e^{-} + O_2 \xrightarrow{k_5} O^{-}_{2(ads)}$$

The concentration of excitons, photoholes and photoelectrons may be considered by applying a steady state treatment:

$$\frac{d[h-e]}{dt} = k_1 I_{(ads)} - k_2[(h-e)] = 0$$

So that:

$$[(h-e)] = \frac{k_1}{k_2} I_{(ads)}$$

Similarly:

$$\frac{d[h]}{dt} = k_2[(h-e)] - k_3[h^{+}][e^{-}] - k_4[h^{+}][OH_{(S)}] = 0$$

Since:

$$[h^{+}] = [e^{-}]$$

Then:

$$\frac{d[h]}{dt} = k_1 I_{(ads)} - k_3\left[h^+\right]^2 - k_4\left[h^+\right]\left[OH^-_{(S)}\right] = 0$$

So that:

$$k_1 I_{(abs)} = k_3\left[h^+\right]^2 + k_4\left[h^+\right]\left[OH^-_{(S)}\right]$$

There are two possibilities concerning the light intensity:

1. At high light intensities, where the recombination of photoholes and photoelectrons is predominate, then:

$$k_3\left[h^+\right]^2 \rangle\rangle k_4\left[h^+\right]\left[OH^-_{(S)}\right]$$

So equation becomes:

$$k_1 I_{(abs)} = k_3\left[h^+\right]^2$$

Then:

$$\left[h^+\right] = \left[\frac{k_1}{k_3}\right]^{\frac{1}{2}} I^{\frac{1}{2}}_{(abs)}$$

If the rate controlling step in the overall photocatalysis processes involves the surface trapping of photoholes at surface OH-, then the rate will be given by equation. The reaction rate is given by:

$$\text{Reaction rate} = k_4\left(\frac{k_1}{k_3}\right)^{\frac{1}{2}} I^{\frac{1}{2}}_{(abs)}\left[OH^-_{(S)}\right]$$

Then:

$$\text{Reaction rate}\,\alpha\left[I_{(abs)}\right]^{\frac{1}{2}}$$

If, on the other hand, the rate controlling step involves photoelectron trapping by oxygen, then the rate controlling step will be equation. The reaction rate is given by:

$$\text{Reaction rate} = k_5\left(\frac{k_1}{k_3}\right)^{\frac{1}{2}} I_{abs}\left[O_{2(abs)}\right]$$

Then:

$$\text{Reaction rate}\,\alpha\left[I_{(abs)}\right]^{\frac{1}{2}}$$

- At low light intensities, it is expected that recombination of photoholes and photoelectrons will be low, then:

$$k_4\left[h^+\right]\left[OH^-_{(S)}\right] \rangle\rangle k_3\left[h^+\right]^2$$

So equation becomes:

$$k_1 I_{(abs)} = k_4\left[h^+\right]\left[OH^-_{(S)}\right]$$

Then:

$$\left[h^{+}\right]=\frac{k_1\left[I_{(abs)}\right]}{k_4\left[OH^{-}_{(S)}\right]}$$

It follows that: $\text{Reaction rate} = k_1\left[I_{(abs)}\right]$

Alternatively, if photoelectron trapping is considered to be rate controlling, then:

$$\text{Reaction rate} = k_5\frac{k_1}{k_4}\left[\left(I_{(abs)}\right)\right]$$

Hence, a linear dependence would be expected at low light intensities. Square-root intensity dependence was observed with rutile I, rutile 11, anatase, uncoated anatase pigment and platinized anatase and also independent on wavelength of incident radiation. Bahnemann *et al.*, (1991) reported that the change in kinetic constant is a function of the square root of the radiation entering at high light intensities, while this change can be linear with light intensity of incident radiation low light intensities. Ollis *et al.*, (1991) summarize the effect of light intensity on the kinetics of the photocatalytic degradation of dye as follows:

- At low light intensities (0–20 mW/cm^2), the rate of photocatalytic degradation is proportional directly with light intensity (first order).
- At high light intensities (25 mW/cm^2), the rate of photocatalytic degradation is proportional directly with the square root of the light intensity (half order).
- At high light intensities the rate of photocatalytic degradation is independent of light intensity (zero order).

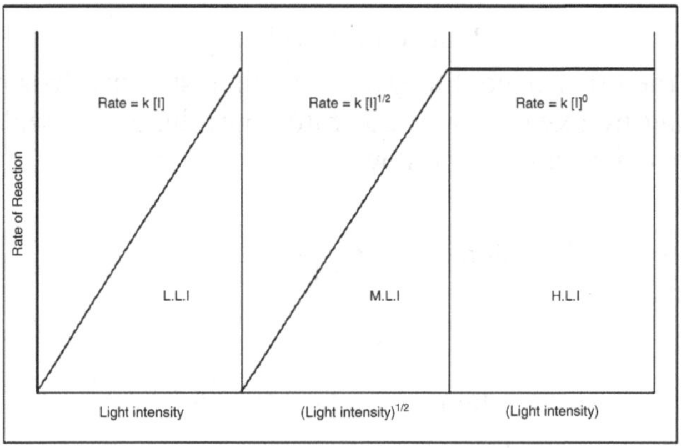

Fig. Effect of Light Intensity on the Kinetics of the Photocatalytic Degradation of Dye

However, Hussein *et al* (2011) found that the rate of photocatalytic decolorization of Bismarck brown R on ZnO and different types of titanium dioxide is proportional directly with the light intensity of incident UVA radiation in the range of 0- 2.0 mW/cm^2 and with the square root of light intensity in the range of

2.0- 3.5 mW/cm^2. Figures illustrate the impact of initial light intensity on the value of rate constant for photocatalytic decolorization of Bismarck brown R on ZnO and different types of titanium dioxide, respectively.

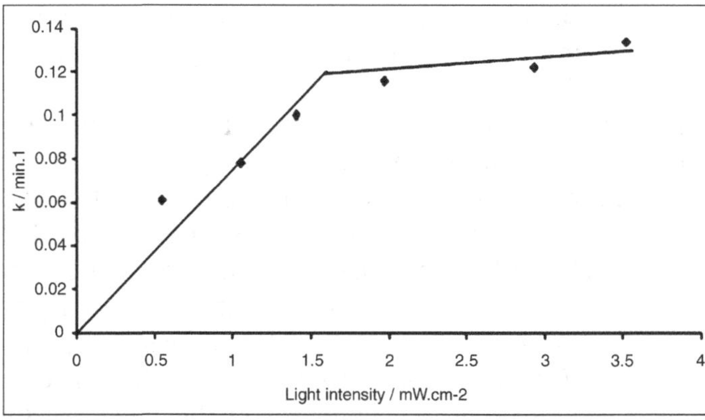

Fig. Effect of Initial Light Intensity on Rate Constant of Photocatalytic Decolorization of Bismarck Brown R on ZnO

The results indicate that the photocatalytic decolorization of Bismarck brown R increases with the increase in light intensity, attaining a maximum value at 3.52 $mW.cm^{-2}$.

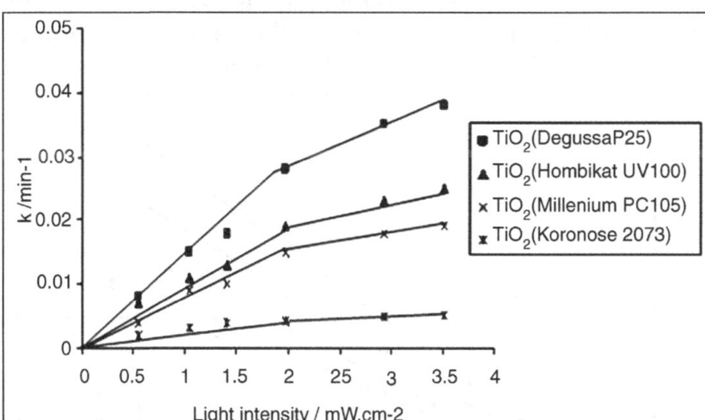

Fig. Effect of Initial Light Intensity on Rate Constant of Photocatalytic Decolorization of Bismarck Brown R using Different Types of TiO_2

EFFECT OF TEMPERATURE

One of the advantages of photoreaction is that it is not affected or slightly affected by temperature change. Temperature dependent steps in photocatalytic reaction are adsorption and desorption of reactants and products on the surface of photocatalyst. None of these steps appears to be rate determining. The impact of temperature is explained as the variable with the least effect on photocatalytic degradation of aqueous solution of azo dyes (Obies, 2011).

Attia *et al.*, (2008) have found that the activation energy of photodegeradation of real textile industrial wastewater is equal to 21 ± 1 kJ mol^{-1} on titanium dioxide and 24 ± 1 kJ mol^{-1} on zinc oxide. The activation energy for the photocatalytic degradation of textile industrial wastewater on titanium dioxide is similar to previous findings for photocatalytic oxidation of different types of alcohols on titanium dioxide and metalized titanium dioxide.

The single value of activation energy (21 ± 1 kJ mol^{-1}) that can be related to the calculated activation energy of photooxidation of different species of titanium oxide is associated with the transport of photoelectron through the catalyst to the adsorbed oxygen on the surface (Harvey *et al.*, 1983 a & b). Kim and Lee, (2010) explained that the very small activation energy in photocatalytic reactions is the apparent activation energy E_a, whereas the true activation energy E_t is nil. These types of reactions are operating at room temperature.Palmer *et al.*, (2002) observed that the effect of temperature on the photocatalytic degradation is insignificant in the range of 10-68°C. High temperatures may have a negative impact on the concentration of dissolved oxygen in the solution and consequently, the recombination of holes and electrons increases at the surface of photocatalyst. However, Trillas *et al.*, (1995); Chen and Ray, (1998) reported that raising the temperature of reaction enhances the rate of photocatalytic degradation significantly. Hussein and Abbas (2010 b) reported that the decolorization efficiency of real textile industrial wastewater increases with increasing of temperature as shown in fig.

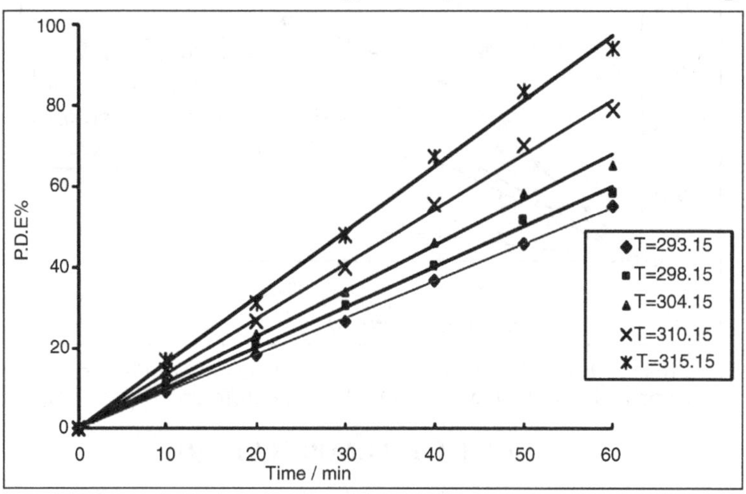

Fig. Effect of Temperature on P.D.E. of Real Textile Industrial Wastewater on Anatase under Solar Radiation

Hussein *et al.*, (2011) have found that the rate of decolorization of Bismarck brown R on ZnO and different types of TiO_2 increases slightly with the increase of the temperature and the activation energy 24 ± 1 kJ.mol^{-1} for ZnO and 14 ± 1, 16 ± 1, 21 ± 1 and 22 ± 1 kJ.mol^{-1} for TiO_2 (Degussa P25), TiO_2 (Hombikat

UV100), TiO_2 (Millennium PC105), and TiO_2 (Koronose 2073), respectively. Figure shows the impact of temperature on photodecolorization of Bismarck brown R by using TiO_2 (Hombikat UV100).

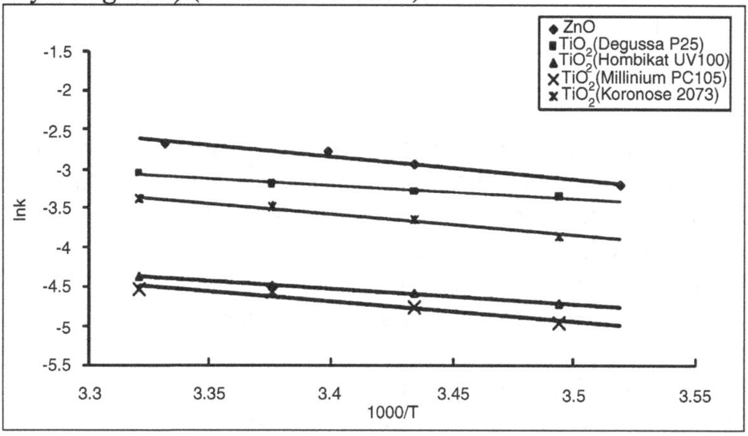

Fig. Arrhenius Plot by Different Types of Catalyst with Bismarck Brown R

EFFECT OF ADDITION OF OXIDANTS

It is well known that the addition of oxidants increases the rate of photocatalytic degradation of dyes by the formation of hydroxyl radicals (Salvador and Decker, 1984; Jenny and Pichat, 1991). However, this is not general for all types of dyes. Production of additional hydroxyl radicals occurs when hydrogen peroxide is added through the following mechanisms (Galvez, 2003 and Dong *et al.*, 2010):

- Trapping of photogenerated electrons.

$$H_2O_2 + 2e^- \rightarrow 2OH^-$$

- Self-decomposition by photolysis

$$H_2O_2 + hv \rightarrow 2\dot{O}H$$

- Reaction with superoxide radical anion $O2^{\bullet-}$

$$H_2O_2 + O_2^- \rightarrow \dot{O}H + OH^- + O_2$$

The addition of persulphate leads to form sulphate radical anion by trapping the photogenerated electrons (Konstantinou and Albanis, 2004)

$$S_2O_8^{2-} + e^- \rightarrow SO_4^{2-} + SO_4^-$$

The formed sulphate radical anion is a strong oxidant and reacts with organic molecules pollutants as follows:

- Abstracting a hydrogen atom from saturated carbon.
- Adding hydrogen to unsaturated or aromatic carbon.

- Removing one electron from carboxylate anions and from certain neutral molecules.

Sulphate radical anion can also react with water molecule to produce hydroxyl radical:

$$SO_4^{\cdot-} + H_2O \rightarrow SO_4^{2-} + \dot{O}H + H^+$$

Other oxidants such as iodate and bromate can also increase the reaction rate because they are also electron scavengers, while chlorate has been proven insufficient to improve effectiveness (Galvez, 2003). However, these additives are too expensive to be compared to hydrogen peroxide and peroxydisulphate. Moreover, they do not dissociate into harmless products.

The addition of oxidant to reaction mixture serves the rate of photocatalytic degradation by:

- Generation of additional •OH and other oxidizing species.
- Increasing the number of trapped photoelectrons.
- Increasing the oxidation rate of intermediate compounds.
- Replacement of oxygen role in the case of the absence of oxygen in the reaction mixture.

Table shows the effect of addition of hydrogen peroxide on the rate of photocatalytic degradation of red disperse dye on ZnO. The results indicate that the apparent rate constant increases with the increase in H_2O_2 concentration to a certain level and a further increase in H_2O_2 concentration leads to decrease in the degradation rate of the red disperse dye.The presumed reason is that the addition of H_2O_2 to a certain level increases the production of hydroxyl radicals, but the additional amount leads to reduce the amounts of photoholes and hydroxyl radicals:

$$H_2O_{2+}2h^+ \rightarrow O_2 + 2H^+$$

$$H_2O_2 + \dot{O}H \rightarrow H_2O + HO_2^{\cdot}$$

$$HO_2^{\cdot} + \dot{O}H \rightarrow H_2O + O_2$$

Table. Effect of Addition of H_2O_2 on the Apparent Rate Constant of Photocatalytic Decolorization of Disperse Dye.

Conc. of H_2O_2 (mol.L^{-1})	K_{app} (min^{-1})
0	0.0893
1.5	0.1408
3.0	0.1499
4.5	0.1680
6.0	0.1404
7.5	0.1290

This behaviour relates to the competition between the adsorption of organic pollutants and H_2O_2 on the surface of photocatalyst. The required amount of H_2O_2 reaches the highest level of enhancement for the rate of photodegradation is related to the ratio of the concentration of organic pollutants and H_2O_2 (Galvez, 2003). When the pollutant concentration is low compared with the concentration of H_2O_2, the adsorption of organic pollutants decreases due to the increase of adsorption of hydrogen peroxide and, as a result, the additional hydroxyl radicals generated by H_2O_2 do not react efficiently.

EFFECT OF IRRADIATION SOURCES

Table summarized the obtained results from the different techniques used for the treatments of textile industrial wastewater and different types of industrial dyes. The results show that the decolorization rate of textile industrial wastewater is faster with solar light than with UV light. The results indicated that solar energy could be effectively used for photocatalytic degradation of pollutants in wastewater.

Table. Effect of Irradiation Sources on Photocatalytic Decolorization of Textile Industrial Wastewater and Different Dyes

Process	Type of Treated Waste or Dye	Source of Irradiation	Type of Catalyst Hours	Time for Complete Mineralization/
Photocatalytic	Textile industrial wastewater	Solar	ZnO	2.7
Photocatalytic	Textile industrial wastewater	Solar	TiO_2	4
Photolysis	Murexide	Solar	–	7.5
Photocatalytic	Murexide	Solar	ZnO	3.8
Photocatalytic	Murexide	Solar	TiO_2	3.3
Photolysis	Thymol blue	Solar	–	7
Photocatalytic	Thymol blue	Solar	TiO_2	2.7
Photocatalytic	Thymol blue	Solar	ZnO	3.3
Photocatalytic	Textile industrial wastewater	Mercury lamp	TiO_2	2.6
Photocatalytic	Textile industrial wastewater	Mercury lamp	ZnO	3
Photocatalytic	Bismarck brown G	Mercury lamp	ZnO	1
Photocatalytic	Bismarck brown R	Mercury lamp	ZnO	0.8
Photocatalytic	Bismarck brown R	Mercury lamp	TiO_2	1.2
Photocatalytic	Textile industrial wastewater	Mercury lamp	TiO_2	3
Photocatalytic	Textile industrial wastewater	Mercury lamp	ZnO	1
Photocatalytic	Textile industrial	Solar	TiO_2	1.8

	wastewater			
Photocatalytic	Textile industrial wastewater	Solar	ZnO	0.33

COMPARISON BETWEEN MINERALIZATION AND PHOTOCATALYTIC DECOLORIZATION

Mineralization of dyes is a process in which dyes are converted completely into its inorganic chemical components (minerals), such as carbon dioxide, water and other species according to the structure of dye. Mineralization of Bismarck brown R was evaluated by analysing total organic carbon (TOC). The results shown in Fig. indicate that photocatalytic decolorization of Bismarck R brown was faster than the decrease of TOC. The results indicate that % TOC reduction was about 73% after 60 minutes of irradiation while the per cent of decolorization achieved 88% for the same period of irradiation.

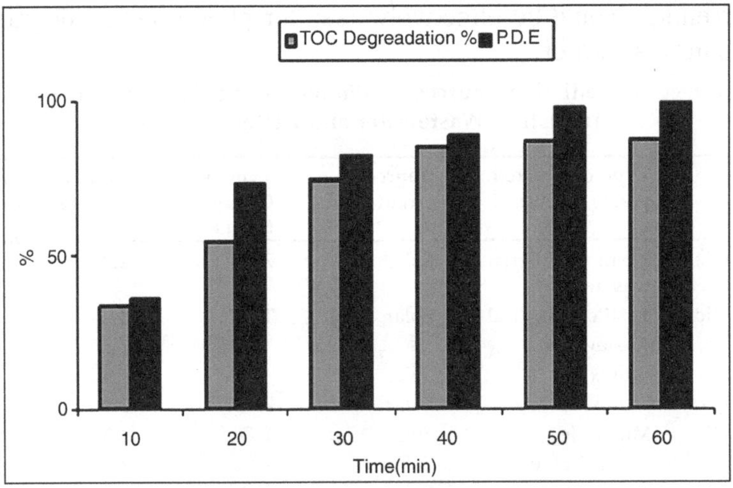

Fig. Comparison between Mineralization and Photocatalytic Decolorization of Bismarck Brown R on ZnO

These findings in good agreement with those reported before. This could be explained to the formation of some by product, which resist the photocatalytic degradation. Furthermore, the formed by products need more time to destruct.

CONCLUSIONS

- Photocatalytic degradation techniques is the most efficient and clean technology.
- Textile industries have become worldwide. Thus, this method can be considered as a promising technique for providing formidable quantities of water especially for countries facing serious suffering from water shortage.

- The existence of catalyst and lights are essential for photocatalytic degradation of colored dyes.
- Photocatalytic degradation efficiency (PDE) of textile industrial wastewater is obviously affected by illumination time, pH, initial dye concentration and photocatalyst loading.
- Solar photocatalytic treatment has been proved to be an efficient technique for decolorization of industrial wastewater through a photocatalytic process and the transformation is practically complete in a reasonable irradiation time.
- In the countries where, intense sunlight is available throughout the year, solar energy could be effectively used for photocatalytic degradation of pollutants in industrial wastewater.
- The zero point charge is 6.4 and 9.0 for TiO_2 and ZnO, respectively above which the surface of photocatalyst is negatively charged by means of adsorbed hydroxyl ions; this favours the formation of hydroxyl radical, and as a result, the photocatalytic degradation of industrial wastewater increases due to inhibition of the photoholes and photoelectrons recombination.

BIODEGRADATION OF TEXTILE WASTEWATER

Considering both the volume and the effluent composition, the textile industry wastewater is rated as the most polluting among all industrial sectors. Important pollutants are present in textile effluents; they are mainly recalcitrant organics, colour, toxicants and inhibitory compounds.

Textile industries however, have caused serious environmental problems because of the wastewater produced. Most textile industries produce wastewater with relatively high BOD, COD, suspended solids and colour. The wastewater may also contain heavy metals depending on the type of coloring substances used. In general, the objective of textile industry wastewater treatment to reduce the level of organic pollutants, heavy metal, suspended solids and colour before discharge into the river.

Coloring substances are used for dyeing and printing processes. The wastewater from these two processes is the most polluted liquid waste in a textile industry. Biological, chemical, physical or the combination of the three treatment technologies can be used to treat textile industry liquid waste (Suwardiyno and Wenten, 2005). It has been proven that some of these dyes and/or products are carcinogens and mutagens (Manu and Chaudhari 2003). A part from the aesthetic deterioration of the natural water bodies, dyes also cause harm to the flora and fauna in the natural environment. So, textile wastewater containing dyes must be treated before their discharge into the environment.

Numerous processes have been proposed for the treatment of coloured waste water *e.g.*, precipitation, flocculation, coagulation, adsorption and wet oxidation. All these methods have different colour removal capabilities, capital costs and operating speed. Among these methods coagulation and adsorption are the commonly used; however, they create huge amounts of sludge which become a pollutant on its own creating disposal problems.

Among low cost, viable alternatives, available for effluent treatment and decolourization, the biological systems are recognised, by their capacity to reduce biochemical oxygen demand (BOD) and chemical oxygen demand (COD) by conventional aerobic biodegradation. Work on the use of combined bacterial process to treat textile wastewater has been carried out over the years by many research groups.

Recent study has used the combination of anaerobic and aerobic steps in an attempt to achieve not only decolourization but also mineralization of dyes. Aerobic processes have been recently used for the treatment of textile wastewater as standalone processes and it is confirmed that they are efficient, costeffective for smaller molecules and that the aerobic reactor is an effective technique to treat industrial wastewater (Coughlin *et al.*,

2002; Coughlin *et al.*, 2003; Buitron *et al.*, 2004; Ge *et al.*, 2004; Sandhaya *et al.*, 2005; Steffan *et al.*, 2005; Sudarjanto *et al.*, 2006). The aerobic reactor has the advantage of being a closed and comparatively homogeneous and stable ecosystem.

Since little is known about this ecosystem, a molecular inventory is the first step to describe this dynamic bacterial community without cultivation. In order to better understand the functions of the bacterial community, a full description of the bacterial ecosystem is required.

Acquisition of DNA sequences is now a fundamental component of most phylogenetic, phylogeographic and molecular ecological studies. Single-strand conformation polymorphism (SSCP) offer a simple, inexpensive and sensitive method for detecting whether or not DNA fragments are identical in sequence, and so can greatly reduce the amount of sequencing necessary (Sunnucks *et al.*, 2000). SSCP can be applied without any a priori information on the species and then can give a more objective view of the bacterial community. SSCP has been applied to study microbial communities from several habits including water, compost and anaerobic digesters.

In this research, we used the mixture design in the experimental design (Minitab 14.0) to optimize the formulation of the predominant strains isolated from textile waste water plant.

After biodegradation, the Chemical Oxygen Demand (COD) and percentage of decolorization were measured. The relationships between the different combinations and products were analysed by Minitab to select the optimal bacterial combination and to investigate the aerobic degradability of a textile industry wastewater in Tunisia by an aerobic Stirred Bed Reactor (SBR).

MATERIALS

The microbial strains were microcapsules of *Sphingomonas paucimobilis* (14×10^7 cfu), *Bacillus sp.* (4.2×10^8 cfu) and *Staphylococcus epidermidis* (7×109 cfu), which were isolated from textile Waste Water plant in KsarHellal, Tunisia. *Sphingomonas paucimobilis*, *Staphylococcus epidermidis* and *Bacillus sp.*

Were isolated in previous works of Ayed *et al.*2009a,b and Ayed *et al* 2010a,b,c with the ability of degrading azo and triphenylmethane dyes (Congo Red, Methyl Red, Methyl Orange, Malachite Green, Phenol Red, Fushin, Methyl Green and Crystal Violet). All chemicals used were of the highest purity available and of analytical grade.

NUTRIENT AGAR PREPARATION

Nutrient agar was used as the growth medium for microbial isolation. For this purpose, 28 g of nutrient agar was dissolved in 1 l of distilled water, and was then autoclaved at 121°C for 20 min. After autoclaving, the agar was left to cool at room temperature for 15 min, and it was then poured out into Sterilin disposable Petri dishes.

MICROBIAL STRAIN

The culture was cultivated and maintained by weekly transfers on to nutrient agar slants. For production experiments, the culture was revived in nutrient broth (pH 7.0) and freshly prepared 3 hold culture (λ600 nm= 1) prepared in Mineral Salt Medium (MSM) at 37°C, 150 rpm (New Brunswick Scientific Shaker, Edison, NJ) was used as the inoculum.

The used medium was composed in 1000 ml of distilled water: glucose (1250 mg/l), yeast extract (3000 mg/l), $MgSO_4$ (100 mg/l); $(NH_4)_2SO_4$ (600 mg/l); NaCl (500 mg/l); K_2HPO_4 (1360 mg/l); $CaCl_2$ (20 mg/l); $MnSO_4$ (1.1 mg/l); $ZnSO_4$ (0.2 mg/l); $CuSO_4$ (0.2 mg/l); $FeSO_4$ (0.14 mg/l) and it was maintained at a constant pH of 7 by the addition of phosphate buffer (Ayed *et al.*, 2010a,b,c).

ACCLIMATIZATION

The acclimatization was performed by gradually exposing *Sphingomonas paucimobilis*, *Bacillus sp.* and *Staphylococcus epidermidis* to the higher concentrations of effluent (Kalme *et al.*, 2006). This bacteria were grown for 24 h at 30°C in 250 ml Erlenmeyer flasks containing in g/l yeast extract (3.0) and glucose (1.25) (pH 7.0).

During the investigation, nutrient broth concentration was decreased from 90% (w/v) to 0% (w/v) and finally the organism was provided with effluent as sole source of nutrient. Acclimatization experiments were carried out at optimum temperature.

OPERATIONAL CONDITIONS OF LABORATORY BIOREACTORS

A laboratory scale aerobic bioprocess was used in this study. The aerobic system used was SBR bioreactor. The system was operated continuously at a constant temperature of 30°C using an external water bath. A continuous stirred tank reactor with a 500 ml working volume was used. Mixing was assured by the continuous rotation of the magnetic stirrer.

The system was first inoculated with a microbial consortia (*Sphingomonas paucimobilis*, *Bacillus sp.* and *Staphylococcus epidermidis*) obtained from a textile wastewater treatment plant. These inocula were selected because of the large variety of microorganisms that could be found in the biomass degrading dyes in textile wastewater, and because mixed cultures offer considerable advantages over the use of pure culture. In fact, individual strains may attack the dye molecules at different position or may use decomposition products produced by another strains for further decomposition.

In fact, it is mentioned that adaptation is important for successful decolorization, and as acclimation occurred, the decolorization time becomes constant (Buitron and Quezada Moreno, 2004). The system was fed by a peristaltic pump with the textile effluent obtained from textile wastewater plant in Ksar Hellal (Tunisia), and its pH was maintained at approximately 7. Air was provided from the bottom of the aeration of the combined bacterial process using diffusers and an air pump. Bioreactors operating conditions were (COD: 1700 (mg O_2/l); BOD_5: 400 (mg O_2/l); Colour: 3600 (U.C); pH: 7; MES: 810 (mg/l)).

ANALYTICAL METHODS

The effluent from each bioreactor was collected daily, centrifuged at 6000 rpm for 10 min and analysed for colour, COD, pH, volatile suspended solids (VSS) and colonies forming units (cfu). COD and colour measurements were carried out on the clear supernatant. Colour was measured by an UV–vis spectrophotometer (Spectro UV-Vis Double Beam PC Scnning spectrophotomètre UVD-2960) at a wavelength of 275 nm in which maximum absorbance spectra was obtained. The decolorization and COD removal were calculated according to the following formulation (Eq 1and Eq 2) (Ayed *et al.*, 2009a,b). In this study, *Sphingomonas paucimobilis*, *Bacillus sp.* and Filamentous bacteria were used as mixture starters, with different proportions ranging from 0 to 100%. Decolorization experiments were taken according to the ratio given by the experimental design, and 10% of mixed culture were inoculated into the effluent (3.0 g/l yeast extract and 1.25 g/l glucose) at 37°C for 10 hin shaking conditions (150 rpm) (Ayed *et al.*, 2010a,b,c).

$$\% \text{ Decolorization} = \frac{(I - F)}{I} \times 100$$

Where I was the initial absorbance and F the absorbance at incubation time t

$$\text{COD removal}(\%) = \frac{\text{initial COD}(0\,\text{h}) - \text{observed COD}(t)}{\text{initial COD}(0\,\text{h})} \times 100$$

The pH was measured using a digital calibrated pH-meter (Inolab, D-82362 Weilheim Germany). All assays were carried in triplicate.

Table. Mixture Design Matrix with the Experimental Analysis

Assay	Sphingomonas Paucimobilis	Bacillus sp.	Staphylococcus Epidermidis	Total	COD Removal (%)	Decolorization (%)
1	0.66667	0.16667	0.16667	1.00000	60	70
2	0.50000	0.50000	0.00000	1.00000	76	77
3	0.50000	0.00000	0.50000	1.00000	77	88
4	0.33333	0.33333	0.33333	1.00000	75	80
5	0.00000	1.00000	0.00000	1.00000	70	77
6	1.00000	0.00000	0.00000	1.00000	81	89
7	0.16667	0.16667	0.66667	1.00000	53	63
8	0.00000	0.00000	1.00000	1.00000	49	55
9	0.00000	0.50000	0.50000	1.00000	45	44
10	0.16667	0.66667	0.16667	1.00000	60	64

PILOT PLANT DESIGN

As described earlier, the pilot plant comprised several treatment steps.

MODEL ESTABLISHMENT

Through linear regression fitting, the regression models of tow responses (COD % and decolorization %) were established. The regression model equations are as follows:

Ydecolorization % = 85.34 S1 + 67.70 S2 + 55.43 S3+ (-21.77) S1*S2+ (67.69) S1*S3+ (-73.59) S2*S3 R^2= 84.82%; P=0.09

YCOD% = 77.11 S1 + 70.11 S2 + 49.02 S3 + (-1.55) (S1*S2) + (44.27) (S1*S3) + (-57.73) (S2*S3) R^2 = 75.27%; P = 0.2

Where S1: Sphingomonas paucimobilis; S2: Bacillus sp. and S3: Staphylococcus epidermidis

EFFECT OF FORMULATION ON THE PERCENTAGE OF DECOLORIZATION AND COD REMOVAL OF EFFLUENT

The mixture design is now used widely in the formulation experiment of food, chemicals, fertilizer, pesticides, and other products. It can estimate the relationship between formulation and performance through regression analysis in fewer experiment times.

Mixture Contour Plot of COD removal (%)

(component amounts)

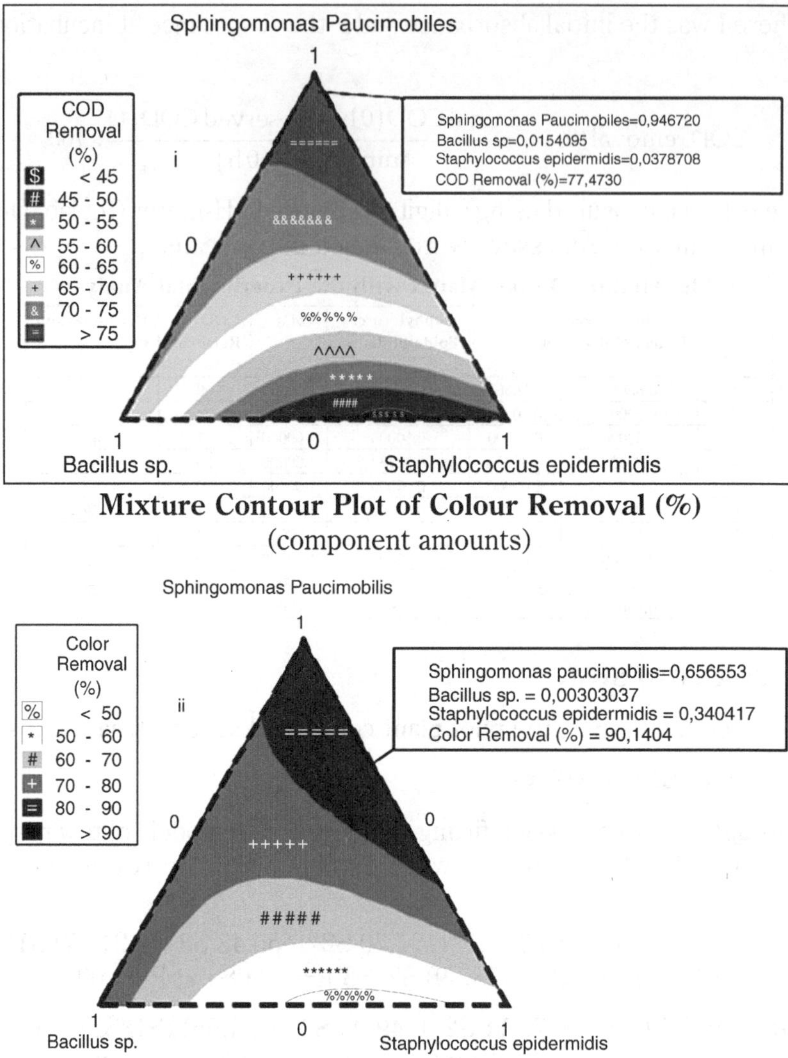

Fig. Mixture Contour Plots between the Variables (*Sphingomonas Paucimobilis, Bacillus sp* and *Staphylococcus Epidermidis.*) for i COD Removal (%), ii Colour Removal (%).

Zhang *et al*. (2006) studied the formulation of plant protein beverage using the mixture design, obtaining the optimized combination of walnut milk, peanut milk, and soy milk. In the mixture design, the effect of the change of variables on the responses can be observed on the ternary contour map.

Figure shows the effect of the interaction of *Sphingomonas paucimobilis, Bacillus sp*. and *Staphylococcus epidermidis* on the decolorization of effluent; Figure shows the effect of the interaction of *Sphingomonas paucimobilis, Bacillus sp*. And *Staphylococcus epidermidis* on the variation of COD.

The statistical significance of the ratio of mean square variation due to regression and mean square residual error was tested using analysis of variance.

ANOVA is a statistical technique which subdivides the total variation in a set of data into component parts associated with specific sources of variation for the purpose of testing hypotheses on the parameters of the model. Only results obtained for decolorization and COD removal were presented herein for clarity of purpose.

The regression adjusted average squares were (305.8) and (231), the linear regression adjusted average squares were (1529.3) and (1115.02) allowed the calculation of the Fisher ratios (*F*-value) for assessing the statistical significance. The model *F* -value (4.33) and (2.43) implies that most of the variation in the response can be explained by the regression equation.

Table. Analysis of Variance of % Decolorization (ANOVA) for the Selected Linear and Interactions Model for Effluent Textile Wastewater

Source	Degrees of Freedom	Sum of Square	Sum of Adjusted Squares	Adjusted Average Squares	*F*-ratio	*P*-value (Significance)
Regression	5	1529,3	1529,303	305,861	4,33	0,090
Linear regression	2	996,33	521,279	260,639	3,69	0,123
Quadratic regression	3	532,97	532,970	177,657	2,52	0,197
Residual error	4	282,30	282,297	70,574		
Total	9	1811,60				

The *P*–value for the regression obtained R2= 84.82%; P=0.09 for decolorization was less than 0.1 and means consequently that at least one of the term in the regression equation has significant correlation with the response variable.

The associated *P*–value is used to judge whether *F*–ratio is large enough to indicate statistical significance.

A *P*-value is more than 0.1 (*i.e.* α =0.05 or 95% confidence) indicates that the model is not to be considered statistically significant. The non-significant value of lack of fit (>0.05) revealed that the quadratic model is statistically significant for the response and therefore it can be used for further analysis. The ANOVA test also shows a term for residual error, which measures the amount of variation in the response data left unexplained by the model (Xudong and Rong, 2008).

The collected data were analysed by using Minitab 14 Statistical Software for the evaluation of the effect of each parameter on the optimization criteria. In order to determine the effective parameters and their confidence levels on the colour removal process, an analysis of variance was performed.

A statistical analysis of variance (ANOVA) was performed to see which process parameters were statistically significant.

Table. Analysis of Variance of COD% (ANOVA) for the Selected Linear and Interactions Model for Effluent Textile Wastewater

Source	Degrees of Freedom	Sum of Square	Sum of Adjusted Squares	Adjusted Average Squares	*F*-ratio	*P*-value (Significance)
Regression	5	1155,02	1155,023	231,005	2,43	0,205
Linear regression	2	885,44	470,406	235,203	2,48	0,199
Quadratic regression	3	269,58	269,578	89,859	0,95	0,498
Residual error	4	379,48	379,477	94,869		
Total	9	1534,50				

F-test is a tool to see which process parameters have a significant effect on the dye removal value. The *F*-value for each process parameter is simply a ratio of the mean of the squared deviations to the mean of the squared error. The colour removal from the real textile wastewater was investigated in different experimental conditions.

The mixture surface plots, which are a three-dimensional graph, was represented using COD and colour removal were represented based on the simultaneous variation of *Sphingomonas paucimobilis*, *Bacillus sp.* and *Staphylococcus epidermidis* in the consortium composition ranging from 0 to 100 % for each strain.

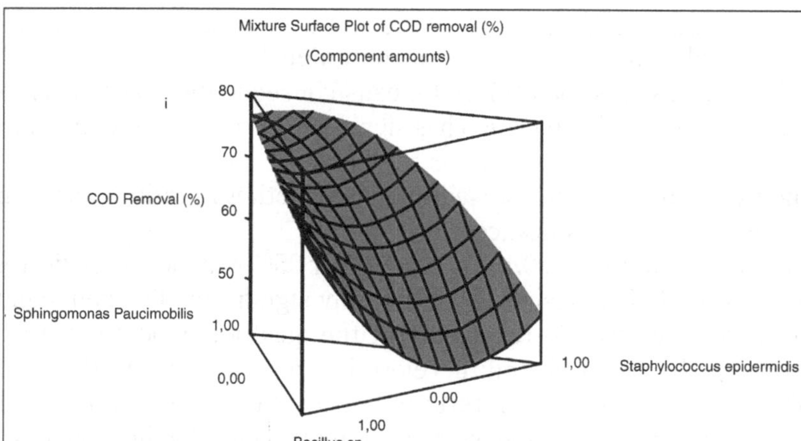

The mixture surface plot also describing individual and cumulative effect of these three variables and their subsequent effect on the response (Liu *et al.*, 2009; Ayed *et al.*, 2010b,c). The mixture contour plots between the variables such as *Sphingomonas paucimobilis*, *Bacillus sp.* and *Staphylococcus epidermidis* are given in Figure. The lines of contour plots predict the values of each response at different proportion of *Sphingomonas paucimobilis*, *Bacillus sp.* and *Staphylococcus epidermidis*. These values are more or less same to the experimental values.

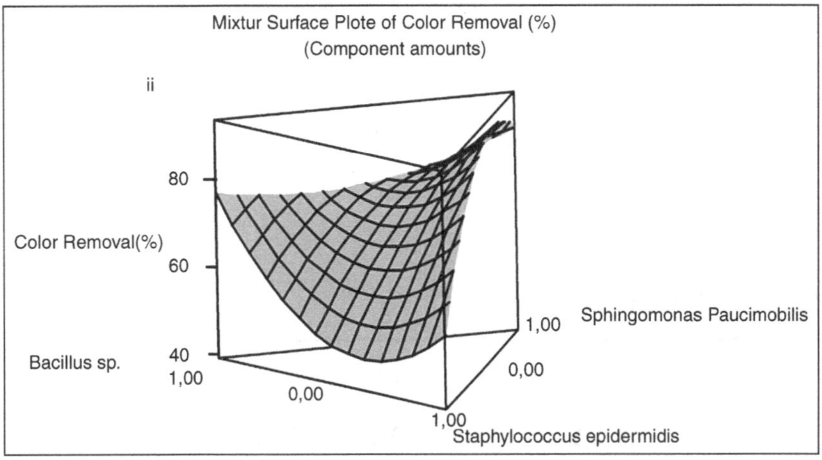

Fig. Mixture Surface Plots between the Variables (*Sphingomonas Paucimobilis, Bacillus sp* and *Staphylococcus Epidermidis.*) for i COD Removal (%), ii Colour Removal (%).

The developed consortium showed a better decolorization yields as compared to pure cultures, which proved a complementary interaction among various isolated bacteria. The consortium achieved significantly a higher reduction in colour (90.14%) and COD removal (77.47%) in less time (96h). The biodegradation of the effluent textile wastewater was achieved by the developed consortium using *Sphingomonas paucimobilis*, *Bacillus sp.* and *Staphylococcus epidermidis*.

8

Dyes Industry

DYES IN FOOD INDUSTRY

A food coloring is any substance that is added to food or drink to change its color. Food coloring is used both in commercial food production and in domestic cooking. Due to its safety and general availability, food coloring is also used in a variety of non-food applications, for example in home craft projects and educational settings

Purpose of Food Coloring

People associate certain colors with certain flavors, and the color of food can influence the perceived flavor, in anything from candy to wine. For this reason, food manufacturers add dyes to their products. Sometimes the aim is to simulate a color that is perceived by the consumer as natural, such as adding red coloring to glacé cherries (which would otherwise be beige), but sometimes it is for effect, like the green ketchup that Heinz launched in 2000.

While most consumers are aware that food with bright or unnatural colors (such as the green ketchup or children's cereals such as Froot Loops) likely contain food coloring, far fewer people know that seemingly "natural" foods such as oranges and salmon are sometimes also dyed to mask natural variations in color. Color variation in foods throughout the seasons and the effects of processing and storage often make color addition commercially advantageous to maintain the color expected or preferred by the consumer. Some of the primary reasons include:

- Offsetting color loss due to light, air, extremes of temperature, moisture, and storage conditions.
- Masking natural variations in color.
- Enhancing naturally occurring colors.
- Providing identity to foods.
- Protecting flavors and vitamins from damage by light.
- Decorative or artistic purposes such as cake icing.

Regulation

Food colorings are tested for safety by various bodies around the world and sometimes different bodies have different views on food color safety. In the United States, FD&C (generally indicates that the FDA has approved the colorant for use in foods, drugs and cosmetics) numbers are given to approved synthetic food dyes that do not exist in nature, while in the European Union, E numbers are used for all additives, both synthetic and natural, that are approved in food applications.

Most other countries have their own regulations and list of food colors which can be used in various applications, including maximum daily intake limits.

Natural colors are not required to be tested by a number of regulatory bodies throughout the world, including the United States FDA.. However, this list contains substances which may have synthetic origins.

NATURAL FOOD DYES

A growing number of natural food dyes are being commercially produced, partly due to consumer concerns surrounding synthetic dyes. Some examples include:

- Caramel coloring, made from caramelized sugar, used in cola products and also in cosmetics.
- Annatto, a reddish-orange dye made from the seed of the Achiote.
- A green dye made from chlorella algae.
- Cochineal, a red dye derived from the cochineal insect, *Dactylopius coccus*.
- Betanin extracted from beets.
- turmeric
- saffron
- paprika
- Elderberry juice

To ensure reproducibility, the colored components of these substances are often provided in highly purified form, and for increased stability and convenience, they can be formulated in suitable carrier materials (solid and liquid).

Caramel coloring

Caramel coloring is a water soluble food coloring; it is made by the carefully controlled heat treatment of carbohydrates, generally in the presence of acids, alkalis, or salts, in a process called caramelization. It is more fully oxidized than caramel candy and has an odor of burnt sugar and a somewhat bitter taste. Its color ranges from pale yellow to amber to dark brown.

Internationally, the Joint FAO/WHO Expert Committee on Food Additives (JECFA) recognizes four classes of caramel coloring, differing by intended application and in the reactants used in their manufacture

- (Class I): *Plain caramel, caustic caramel, spirit caramel*, INS No. 150a, or E150a; Prepared by heating carbohydrates with or without acids or alkalis but no ammonium or sulfite compounds can be used.
- (Class II): *Caustic sulfite caramel*, INS No. 150b, or E150b; Prepared by heating carbohydrates with or without acids or alkalis in the presence of sulfite compounds but no ammonium compounds are used.
- (Class III): *Ammonia caramel, baker's caramel, confectioner's caramel, beer caramel* INS No. 150c, or E150c; Prepared by heating carbohydrates with or without acids or alkalis in the presence of ammonium compounds but no sulfite compounds are used. Used in beer, synthetic soy sauce, and confectionery.
- (Class IV): *Sulfite ammonia caramel, acid-proof caramel, soft-drink caramel*, INS No. 150d, or E150d; Prepared by heating carbohydrates with or without acids or alkalis in the presence of both sulfite and ammonium compounds. Used in acid environments such as soft drinks.

Caramel coloring molecules carry either a positive or negative charge depending upon the reactants used in their manufacture. Problems such as precipitation, flocculation or migration can be eliminated with the use of a properly charged caramel coloring for the intended application.

Uses

Caramel coloring is one of the oldest and most widely-used food colorings, and is found in almost every kind of industrially produced food, including: beer, brown bread, buns, chocolate, cookies, brandy, chocolate flavored flour-based confectionery, coatings, decorations, fillings and toppings, chips, dessert mixes, doughnuts, fish and shellfish spreads, frozen desserts, glucose tablets, cough drops, gravy browning, ice cream, jams, milk desserts, pancakes, pickles, sauces and dressings, soft drinks (especially colas), stouts, sweets, vinegar, whiskey, and wines.

Production

Caramel coloring is made by the controlled heat treatment of carbohydrates (nutritive sweeteners which are the monomers glucose and fructose or their polymers, *e.g.* glucose syrups, sucrose, invert syrups, and dextrose, generally in the presence of food-grade acids, alkalis, and/or salts, in a process called caramelization. Antifoaming agents may be used as processing aids during its manufacture. Unlike caramel candy, it tends towards maximum oxidation of the sugar to produce a caramel concentrate that has an odor of burnt sugar and a somewhat bitter taste in its raw liquid form. Its color ranges from pale yellow to amber to dark brown.

Annatto

Annatto, sometimes called *Roucou*, is a derivative of the achiote trees of tropical regions of the Americas, used to produce a red food coloring and also as a flavoring. Its scent is described as "slightly peppery with a hint of nutmeg" and flavor as "slightly sweet and peppery".

Annatto is produced from the reddish pulp which surrounds the seed of the achiote (*Bixa orellana* L.). It is used in many cheeses (e.g., Cheddar, Red Leicester, and Brie), margarine, butter, rice, smoked fish and custard powder.

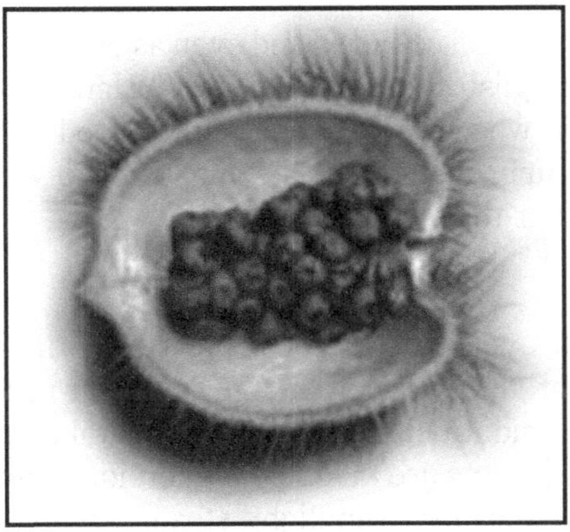

Bixa Orellana seeds.

Annatto is commonly found in Latin America and Caribbean cuisines as both a coloring agent and for flavoring. Central and South American natives use the seeds to make a body paint, and lipstick. For this reason, the achiote is sometimes called the *lipstick-tree*.

Uses

Annatto has had many uses over the centuries, including as a food dye, body paint, treatment for heartburn and stomach distress, sunscreen and insect repellent. In Venezuela, annatto (called locally *onoto*) is used in the preparation of hallacas, perico, and other traditional dishes. In Brazil, both annatto (the product) and the tree (*Bixa orellana* L.) are called *urucum* and the product itself may also be called *colorau*. In the Caribbean islands, both fruit and tree are popularly called *achiote* or *bija* (pronounced "bee-ha") instead of Bixa. In the Philippines, it is called *atsuete*[11] and is used as food coloring in traditional dishes. It is a major ingredient in the popular spice blend "Sazón" made by Goya Foods

USE AS A FOOD COLORING

As a food additive, annatto has the E number E160b. The fat soluble part of the crude extract is called bixin, the water soluble part is called norbixin, and both share the same E number as annatto. In the United States, annatto extract is listed as a color additive "exempt from certification and is commonly considered to be a natural color. The yellowish orange color is produced by the compounds bixin and norbixin, which are classified as xanthophylls, a type of carotenoid. However, unlike beta-carotene, another well-known carotenoid, they do not have the correct chemical structures to be vitamin A precursors The more norbixin in an annatto color, the more yellow it is; a higher level of bixin gives it a more reddish shade. Unless an acid-proof version is used, it takes on a pink shade at low pH.

Cheddar cheese is often coloured and even as early as 1860 the real reason for this was unclear: English cheesemaker Joseph Harding stated "to the cheese consumers of London who prefer an adulterated food to that which is pure I have to announce an improvement in the annatto with which they compel the cheesemakers to colour the cheese"

Cochineal

A deep crimson dye is extracted from the female cochineal insects. Cochineal is used to produce scarlet, orange and other red tints. The colouring comes from carminic acid. Cochineal extract's natural carminic-acid content is usually 19–22%. The insects are killed by immersion in hot water (after which they are dried) or by exposure to sunlight, steam, or the heat of an oven. Each method produces a different colour which results in the varied appearance of commercial cochineal. The insects must be dried to about 30 percent of their original body weight before they can be stored without decaying. It takes about 155,000 insects to make one kilogram of cochineal.

There are two principal forms of cochineal dye: *cochineal extract* is a colouring made from the raw dried and pulverised bodies of insects, and *carmine* is a more purified colouring made from the cochineal. To prepare carmine, the powdered insect bodies are boiled in ammonia or a sodium carbonate solution, the insoluble matter is removed by filtering, and alum is added to the clear salt solution of carminic acid to precipitate the red aluminium salt. Purity of colour is ensured by the absence of iron. Stannous chloride, citric acid, borax, or gelatin may be added to regulate the formation of the precipitate. For shades of purple, lime is added to the alum.[2]

As of 2005, Peru produced 200 tonnes of cochineal dye per year and the Canary Islands produced 20 tonnes per year.[10][5] Chile and Mexico have also recently begun to export cochineal.[1] France is believed to be the world's largest importer of cochineal; Japan and Italy also import the insect. Much of these imports are processed and reexported to other developed economies.[10] As of 2005, the market price of cochineal was between 50 and 80 USD per kilogram, while synthetic raw food dyes are available at prices as low as 10–20 USD per kilogram.

Usage

Traditionally cochineal was used for colouring fabrics. During the colonial period, with the introduction of sheep to Latin America, the use of cochineal increased, as it provided the most intense colour and it set more firmly on woolen garments than on clothes made of materials of pre-Hispanic origin such as cotton, agave fibers and yucca fibers.

Once the European market had discovered the qualities of this product, their demand for it increased dramatically, and by the start of the seventeenth century it was traded internationally. Carmine became strong competition for other colourants such as madder root, kermes,

Polish cochineal, brazilwood, and Tyrian purple, as they were used for dyeing the clothes of kings, nobles and the clergy. For the past several centuries it was the most important insect dye used in the production of hand-woven oriental rugs. It was also used for painting, handicrafts, and tapestries. Cochineal-coloured wool and cotton are still important materials for Mexican folk art and crafts.

Now it is used as a fabric and cosmetics dye and as a natural food colouring, as well as for oil paints, pigments and watercolours. When used as a food additive the dye must be included on packaging labels. Sometimes carmine is labelled as *E120*. An unknown percentage of people have been found to have allergies to carmine, ranging from mild cases of hives to atrial fibrillation and anaphylactic shock. Carmine has been found to cause asthma in some people.

Cochineal is one of the colours that the Hyperactive Children's Support Group recommends be eliminated from the diet of hyperactive children. Natural carmine dye used in food and cosmetics can render it unacceptable to vegetarian (or vegan) consumers, and many Muslims and Jews consider carmine-containing food forbidden (haraam and treif) because the dye is extracted from insects.

Cochineal is one of the few water-soluble colourants that resist degradation with time. It is one of the most light- and heat-stable and oxidation-resistant of all the natural colourants and is even more stable than many synthetic food colours.

The water-soluble form is used in alcoholic drinks with calcium carmine; the insoluble form is used in a wider variety of products. Together with ammonium carmine they can be found in meat, sausages, processed poultry products (meat products cannot be coloured in the United States unless they are labeled as such), surimi, marinades, alcoholic drinks, bakery products and toppings, cookies, desserts, icings, pie fillings, jams, preserves, gelatin desserts, juice beverages, varieties of cheddar cheese and other dairy products, sauces, and sweets.

The average human consumes one to two drops of carminic acid each year with food. Carmine is one of the very few pigments considered safe enough for use in eye cosmetics. A significant proportion of the insoluble carmine pigment produced is used in the cosmetics industry for hair- and skin-care products, lipsticks, face powders, rouges, and blushes. A bright red dye and the stain carmine used in microbiology is often made from the carmine extract, too. The pharmaceutical industry uses cochineal to colour pills and ointments

Betanin

Betanin, or Beetroot Red, is a red glycosidic food dye obtained from beets; its aglycone, obtained by hydrolyzing away the glucose molecule, is betanidin. As a food additive, its E number is E162. Betanin degrades when subjected to light, heat, and oxygen; therefore, it is used in frozen products, products with short shelf life, or products sold in dry state.[1] Betanin can survive pasteurization when in products with high sugar content.

Its sensitivity to oxygen is highest in products with high content of water and/or containing metal cations (eg. iron and copper); antioxidants like ascorbic acid and sequestrants can slow this process down, together with suitable packaging.

In dry form betanin is stable in presence of oxygen. Betanin is usually obtained from the extract of beet juice; the concentration of betanin in red beet can reach 300-600 mg/kg. Other dietary sources of betanin and other betalains include the opuntia cactus, Swiss chard, and the leaves of some strains of amaranth.

The color of betanin depends on pH; between four and five it is bright bluish-red, becoming blue-violet as the pH increases. Once the pH reaches alkaline levels betanin hydrolyzes to a yellow-brown color. Betanin can be also used for coloring meat and sausages.

The most common uses of betanin are in coloring ice cream and powdered soft drink beverages; other uses are in some sugar confectionery, e.g. fondants, sugar strands, sugar coatings, and fruit or cream fillings. In hot processed candies, it can be used if added at the final part of the processing. Betanin is also used in soups as well as tomato and bacon products. Betanin absorbs well from the gut and acts as an antioxidant. Betanin is a betalain pigment, together with isobetanin, probetanin, and neobetanin. Other pigments contained in beet are indicaxanthin and vulgaxanthins.

Turmeric

Turmeric (*Curcuma longa*) is a rhizomatous herbaceous perennial plant of the ginger family, Zingiberaceae. It is native to tropical South Asia and needs temperatures between 20° C and 30° C, and a considerable amount of annual rainfall to thrive. Plants are gathered annually for their rhizomes, and re-seeded from some of those rhizomes in the following season. The rhizomes are boiled for several hours and then dried in hot ovens, after which they are ground into a deep orange-yellow powder commonly used as a spice in curries and other South Asian and Middle Eastern cuisine, for dyeing, and to impart color to mustard condiments.

Its active ingredient is curcumin and it has an earthy, bitter, peppery flavor and a mustardy smell. In medieval Europe, turmeric became known as Indian Saffron, since it is widely used as an alternative to the far more expensive saffron spice. Turmeric makes a poor fabric dye as it is not very lightfast (the degree to which a dye resists fading due to light exposure). However, turmeric is commonly used in Indian clothing, such as saris.

Saffron

Saffron's aroma is often described by connoisseurs as reminiscent of metallic honey with grassy or hay-like notes, while its taste has also been noted as hay-like and somewhat bitter. Saffron also contributes a luminous yellow-orange colouring to foods. Confectionaries and liquors also often include saffron. Common saffron substitutes include safflower (*Carthamus tinctorius*, which is often sold as

"Portuguese saffron" or "assafroa") and turmeric (*Curcuma longa*). Medicinally, saffron has a long history as part of traditional healing; modern medicine has also discovered saffron as having anticarcinogenic (cancer-suppressing), anti-mutagenic (mutation-preventing), immunomodulating, and antioxidant-like properties.[64][17][65] Saffron has also been used as a fabric dye, particularly in China and India, and in perfumery.

Most saffron is grown in a belt of land ranging from the Mediterranean in the west to Kashmir in the east. Annually, around 300tonnes of saffron are produced worldwide.

Iran ranks first in the world production of saffron, with more than 94 percent of the world yield. Iran's annual saffron production is expected to hit 300tons by the end of the nation's Fourth Five-Year Socioeconomic Development Plan in 2009. Other minor producers of saffron are Spain, Kashmir, Greece, Azerbaijan, Morocco, and Italy.

A pound of dry saffron (0.45kg) requires 50,000–75,000 flowers, the equivalent of a football field's area of cultivation. Some forty hours of labour are needed to pick 150,000 flowers. Upon extraction, stigmas are dried quickly and (preferably) sealed in airtight containers.

Saffron prices at wholesale and retail rates range from US$500/pound to US$5,000/pound (US$1,100–US$11,000 per kilogram)—equivalent to £250/350 per pound or £5,500/7,500 per kilo. In Western countries, the average retail price is $1,000/£500/700 per pound (US$2,200/£1,100/1,550 per kilogram). A pound comprises between 70,000 and 200,000 threads. Vivid crimson colouring, slight moistness, elasticity, and lack of broken-off thread debris are all traits of fresh saffron.

Paprika

Paprika is a spice made from the grinding of many dried sweet red or green bell peppers (*Capsicum annuum*). In many European countries, the word *paprika* also refers to bell peppers themselves. The seasoning is used in many cuisines to add color and flavor to dishes. Flavors can also vary widely from country to country (ex. Hungarian paprika is traditionally either spicy or sweet).

Usage

Paprika is used as an ingredient in a broad variety of dishes throughout the world. Paprika (*pimentón* in Spain, *colorau* in Portugal, *chiltoma* in Nicaragua, but these "paprikas" are not made exclusively from bell peppers, other varieties are used, and there are several hot and sweet "paprikas") is principally used to season and colour rices, stews, and soups, such as goulash. paprika is also used in the preparation of sausages as an ingredient that is mixed with meats and other spices. Paprika may be smoked for additional flavours, or tastes

Elderberry

Sambucus (elder or elderberry) is a genus of between 5 and 30 species of shrubs or small trees, formerly placed in the honeysuckle family, Caprifoliaceae, but now shown by genetic evidence to be correctly classified in the moschatel family, Adoxaceae. Two of its species are herbaceous.

The genus is native in temperate-to-subtropical regions of both the Northern Hemisphere and the Southern Hemisphere. It is more widespread in the Northern Hemisphere; its Southern Hemisphere occurrence is restricted to parts of Australasia and South America. The leaves are pinnate with 5–9 leaflets (rarely 3 or 11). Each leaf is 5–30cm long, and the leaflets have serrated margins. They bear large clusters of small white or cream-coloured flowers in late spring; these are followed by clusters of small black, blue-black, or red berries (rarely yellow or white).

Uses

The flowers of *Sambucus nigra* are used to produce elderflower liqueur. Wines and cordials may be produced from the berries. Ornamental varieties of *Sambucus* are grown in gardens. Hollowed elderberry twigs have traditionally been used as spiles to tap maple trees for syrup.

DYES AND LAKES

Color additives are available for use in food as either "dyes" or "lakes".

Dyes dissolve in water, but are not soluble in oil. Dyes are manufactured as powders, granules, liquids or other special purpose forms. They can be used in beverages, dry mixes, baked goods, confections, dairy products, pet foods and a variety of other products. Dyes also have side effects which lakes do not, including the fact that large amounts of dyes ingested can color stools. Lakes are the combination of dyes and insoluble material. Lakes tint by dispersion. Lakes are not oil soluble, but are oil dispersible. Lakes are more stable than dyes and are ideal for coloring products containing fats and oils or items lacking sufficient moisture to dissolve dyes. Typical uses include coated tablets, cake and donut mixes, hard candies and chewing gums, lipsticks, soaps, shampoos, talc, etc.

Other Uses

Because food dyes are generally safer to use than normal artistic dyes and pigments, some artists have used food coloring as a means of making pictures, especially in forms such as body-painting. Food colorings can be used to dye fabric, but are usually not washfast when used on cotton, hemp and other plant fibres. Some food dyes can be fixed on Nylon and animal fibers. Red food dye is often used as theatrical blood.

DYES IN LEATHER INDUSTRY

Leather is a material created through the tanning of hides and skins of animals, primarily cattlehide. The tanning process converts the putrescible skin into a durable, long-lasting and versatile natural material for various uses. Together with wood, leather formed the basis of much ancient technology. The leather industry and the fur industry are distinct industries that are differentiated by the importance of their raw materials.

In the leather industry the raw materials are by-products of the meat industry, with the meat having higher value than the skin. The fur industry uses raw materials that are higher in value than the meat and hence the meat is classified as a by-product. Taxidermy also makes use of the skin of animals, but generally the head and part of the back are used. Hides and skins are also used in the manufacture of glue and gelatin.

Leather Dyeing

Leather dyeing usually involves the use of spirit or alcohol based dyes where alcohol quickly gets absorbed into moistened leather, carrying the pigment deep into the surface. “Hi-liters” and “Antiquing” stains can be used to add more definition to patterns.

These have pigments that will break away from the higher points of a tooled piece and so pooling in the background areas give nice contrasts. Leaving parts unstained also provides a type of contrast.

Alternatives to spirit stains might include a number of options. Shoe polish can be used to dye and preserve leather. Oils such as neatsfoot or linseed can be applied to preserve leather but darkens them. A wax paste more often than not serves as the final coat. Sweat and grime will also stain and ‘antique’ leather over time. Gun holsters, saddlebags, wallets and cantanas used by cowboys and buccaroos were rarely colored in the Old West. The red, brown, and black tones develop naturally through handling and as the oiled leathers absorb the rays of the desert sun.

Due to changing environmental laws, alcohol-based dyes are soon to be unavailable. There are currently water-based alternatives available, although they tend not to work as well.

The types of dyestuff that are used by tanneries generally varies depending on the product range needed along with the dictates of the fashion world. It is a fact that each tannery uses between 50-100 or more different types of dyestuffs.

The leather industry primarily uses anionic dyestuffs, that can be divided in the following families:

- Acid dyes
- Direct dyes
- Mordant dyes
- Pre-metalled dyes
- Solubilised sulphur dyes
- Reactive dyes although a part of the anionic group of dyes but use limited to a certain extent.
- Basic Dyes, its use is also limited.

If we see from the chemical point of view the dyes used are mostly Azo dyes, Anthraquinone dyes and Triphenylmethanedyes. Commercially speaking, popular Black dyes account for nearly 60-70% of all leather dyeing. Also, out of this a dye variety called Acid Black 210 dominates with 80-90% usage of all of the black dyes used.

Latest in Leather Dye: Development of High Fastness Dyes

Applying high temperatures releases the fat-soluble components of conventional leather dyes out of their chemical bonds. Now these components has a tendency to migrate upwards into the layers whose task is to give the leather surface finish.

This causes a visible change of color and intensity. What happens is that with the rising of humidity, it is quite likely that non-permanent dyes that are water-soluble, and low in molecular weight, will be washed out. Along with the problems like spotting and discoloration.

Also there is the risk that migrating dye components can rub off on other materials such as plastics or textiles. In comes the latest technology.

The development and production of high-fastness dyes with no migratory secondary components is therefore a central factor in the manufacture of dyes for top quality leather products. The flow chart below describes the process criteria of selecting this dyes.

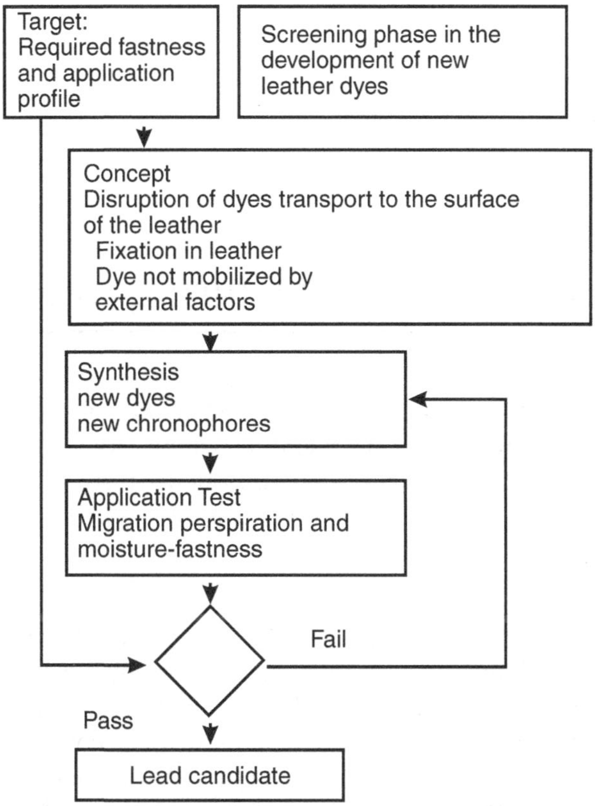

Criterias for determining the Colour Fastness of Dyed Leather

Following are the few important tests that can be undertaken for testing the colour fastness of dyed Leather:-

- Grey scale test to ascertain the change of colour and staining
- Fastness to light the standard Xenon lamp test
- Fastness towards mild washing
- Fastness towards washing by machine
- Colour Fastness with regard to migration into plasticized Poly Vinyl Chloride
- Fastness towards perspiration
- Fastness to test of rubbing
- Fastness in water
- Fastness with respect to water spotting
- Solubility capacity of powder dyes
- Stabilty of dyes with regard to:
  - Acid
  - Alkali
  - Hard water

9

Jute and Ramie Manufacturing in Mills

Raw jute in the form of bales are processed in jute mills to produce hessian, sacking, jute yarn, bags, and other useful products. Raw jute bales from jute fields or suppliers, carried by trucks are unloaded are stacked in the jute mills gowdown. The production of jute goods from raw jute processing involves the following steps:

Selection

In the selection process, raw jute bales are opened to find out any defect and to remove the defective portion from the mora by experienced workers. Raw jute bales are of two types i.e. 150 kg weight and 180 kg weight with or without top portion cutting. The bales are assorted according to end use like Hessiean weft, Sacking wrap, Sacking weft etc. After selection, jute bales are carried to softning section by workers called Gariwala and Bajawala.

Softning

In softning process jute morahs are made soft and pileable. Two methods are used for softning; use of softning machine and use of jute good spreader. Generally an emulsion plant with jute softner machine is used to lubricate and soften the bark and gummy raw jute. The emulsion plant consists of gear pump, motor, vat, jet sprayer, nozzles, emulsion tank and the jacket. In this softning process jute becomes soft and pileable and suitable for carding.

Piling and pile Breaking

The main function of pile breaker is to break the pile and serve it to the carding machines. The softner machine out put material carried by pilemen through a bile to the pile place for pilling. During piling a superficial moisture penetrates inside fibre and "Thermo fillic" action take place which softner the hard portion of the root. After piling for nearly 24 hours the pile breakers carry the material to the carding machine. Generally root cutting is done after piling near the hand feed breaker carding machine. The root weight varies from 5 to 7% of the total weight of jute.

Carding

Carding is a combining operation where jute reeds are splitted and extraneous matters are removed. Jute fibres are formed into ribbon called "sliver". There are three different carding sections:

- Breaker carding
- Inner carding
- Finisher carding

Breaker Carding

In different jute mills the carding operations has been carried out in two ways:

- Hand feed breaker carding
- Rool feed breaker carding

The material after piling more than 24 hours is used in hand feed breaker where the material after piling for 12 hours used in the rool feed carding.

In the Breaker carding machine soften jute after piling is feed by hand in suitable weight. The machine by action with different rollers turns out raw jute in the form of jute sliver for finisher carding. In this process root cutting is necessary before feeding the material to the hand feed breaker carding machine.

Finisher Carding

Finisher carding machine make the sliver more uniform and regular in length and weight obtained from the Breaker carding machine. Finisher carding machine is identical to the Breaker carding machine, having more pair of rollers, staves, pinning arrangement and speed. Nearly 4 to 12 slivers obtained from Breaker carding machine is fed on this machine. The material thus obtained is send to drawing section.

Drawing

Drawing is a process for reducing sliver width and thickness by simultaneously mixing 4 to 6 sliver together. There are three types of Drawing Frame machine. In most mills 3 Drawing passages are used in Hessian and 2 Drawing passages are used in Sacking.

First Drawing

The slivers obtained from finisher carding machine is fed with four slivers on to the first drawing frame machine.The first drawing frame machines makes blending, equalising the sliver and doubling two or more slivers, level and provide quality and colour. This machines includes delivery roller, pressing roller, retaining roller, faller screw sliders, check spring, back spring, crimpling box etc..

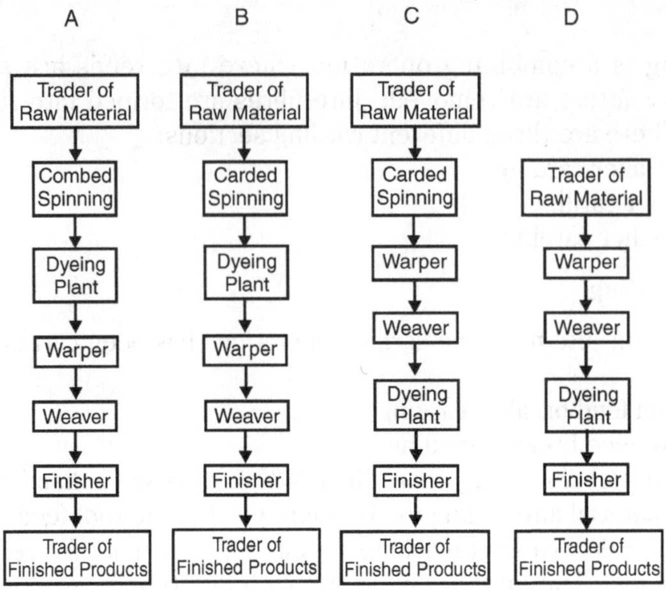

Fig. Process of Finisher Carding

Second Drawing

In second drawing, the Second Drawing Frame machine obtain the sliver from the First drawing machine and use six slivers and deliveries per head. The Second Drawing machine makes more uniform sliver and reduce the jute into a suitable size for third drawing.

Third Drawing

In the third drawing, the Third Drawing frame machine uses the sliver from second drawing. The Third Drawing machine is of high speed makes the sliver more crimpled and suitable for spinning. The comparison of the three drawing process:

Drawing Process	Efficiency Range (%)	Productivity Mt/mc/shift
Ist Drawing	55 – 73	1.75 – 2.2
IInd Drawing	64 – 74	1.62 – 1.9
IIIrd Drawing	67 – 70	1.31 – 1.4

Spinning

Spinning is the process for producing yarn from sliver obtained from Third drawing. In the spinning process slivers are elongated and fibres are twisted into yarn to impart strength. spun yarns in the spinning process are wound onto Bobbins. after to fill the empty bobbins with yarn machine is stopped, replace these bobbins by empty bobbins. The entire time is called an average cycle time. The time for replacing the bobbins full of yarn by empty bobbins is called softing time.

The jute spinning frame machine is fitted with slip draft zone and capable of producing quality yarns at high efficiency with auto-dofting arrangements also. A 4', pithch slip-draft sliver frames available of 20 spindles 100 spindles, having a production range 8 uls to 28 uls with a flyer speed of 3200 to 4000 RPM. Spinning of several types of yarn is processed by spinning frame machine usingdifferent kinds of bobbins, such as: Food Grade HCF, Sacking Wrap, Hessian Wrap, Hessian Weft.

Winding

Winding is a process which provides yarn as spools and cops for the requirement of beaming and weaving operations. There are two types of winding:

- Spool Winding
- Cope Winding

Spool Winding

In Spool Winding yarn is produces for warp (the longitudinal yarn). Spool winding machine consists of a number of spindles. There is wide variation in the number of spindles per machines from one make to another. Productivity of spool winding depends on the surface speed of the spindle and machine utilisation.

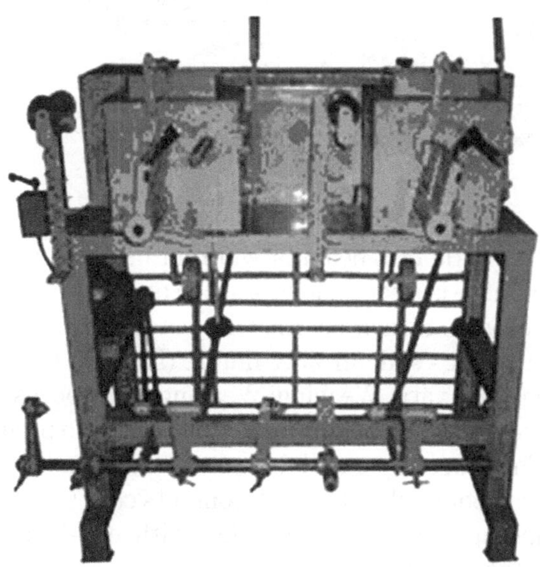

Fig. Spool Winding Machine

Spool winding machine uses the bobbins contain smaller length of yarn. This machine wound the yarn into bigger packages known as 'spool'. The Spool are used in making sheets of yarn to form warp portion used during interleecment of weaving.

Cop Winding

Cop Winding machine obtain yarns from the spinning machines. The spinning bobbins is placed on a suitable pin on top of the cop machine and yarn tension is maintained by means of a small leaver. The yarn on the bobbins are cnverted into hollow cylindrical package said to be cop. The cop is used to form Transverse thread during interlacement of weaving. Generally a cop winding machines consist 120 spindles.

Beaming

Beaming process is follows after spool winding. In Beaming operation yarn from spool is wounded over a beam of proper width and correct number of ends to weave jute cloth. To increase the quality of woven cloth and weaving efficiency, the wrap yarns are coated with starch paste. Adequate moisture is essential in this process.

Fig. Cop Winding Machine

Quality characteristic of a beam is width of beam - number of ends and weight of stand and there is a continuous passage of yarn through starch solution from spools to the beam.Strach solution in water contains tamerine kernel powder (TKP), antiseptic - sodium silica fluride (NaSiF4) and its concentration varies with the quality of yarn. Weaving: Weaving is a process of interlacement of two series of threads called "wrap" and "weft" yarns to produce the fabric of desired quality. There are separate looms for hessian and sacking in weaving section. The Hessian looms, shuttle which contents cops (weft yarn) is manually changed. The sacking looms are equipped with eco-loader to load a cop automatically into the shuttle.

Dumping

Dumping is the process in which the rolled woven cloth is unrolled and water is sprinkled on it continuously to provide desired moisture. Each roll is generally104 yards. Damping is done manually.

Calendering

Calendering is a process similar to ironing of fabric. After damping the damped fabric passes through pairs of heavy rollers rendering threads in fabric flattened and improve the quality and appearance.

Lapping

Lapping is the process in which Hessian fabrics are folded into the required size used in "Bale press" operation on the lapping machine.

Cutting

Cutting is the process where the sacking cloth is cut to the required length for making bags for different size such as A-Twill bags and B-twill bags of 100 kg capacity.

Hemming

In Hemming process, the raw edges of sacking cloth cut pieces are shown by folding it with sewing machine.

Herackele Sewing

In Herackele sewing the sides of sacking cloth cut pieces are shown to make a complete bag.

Bailing

Bags or Bale processing cloths are pressed compactly according to buyers need.

RICE BRAN OIL (RBO) TECHNOLOGY

Rice Bran oil (RBO) Technology is an eco-friendly and the best technology for the production of the safest packaging material for food products. Indian Jute Industries Research Association (IJIRA) and Indian Jute Mills Association (IJMA) initiated research to find a suitable alternative oil for foood grade bags. IJIRA developed a special technology based on Rice Bran oil (RBO II) to manufacture Food Grade Jute Bags. The main feature of Rice Bran Oil Technology are:

- Bio-degradable
- Eco-friendly
- Nontoxic vegetable oil
- Stable and resistant to Rancidity

The bags and cloths processed under this technology fully conform the international specification as prescribed in IJO 98/01 and by IOCCC. The specification for jute packages of food grade material have been accepted by ICCO and recommended by IOCC. The IJO standard 98/01 has three main criterion for food grade bags:

Analytical criteria

The ingradients used as bataching oils shall be nontoxic and shall not contain compound that could produce off-flavours or off-tastes in food materials packed in jute bags.Advance Jute Testing Equipment

Chemical Criteria

Research in Jute Related Fields. The amount of unraponificable compounds shall be les than 1250 mg/kg as recommended by British Standard 3845:1990 subsequent saponification shall be done according to the methodology described the IUPAC 2, 401 of the International Union of Pure and Applied Chemistry.

Organoleptic Criteria

No undesirable odours or odours untypical of jute shall be present. No unacceptable odours shall develop after artificial agency of the bags. The european standard EN 766 on bags for transport of food aid shall be followed.

The above three criteria and other specification are conformed by the food grade bags from India under IBO technology. The quality of Indian food grade bags and cloths is fully insured by IJIRA's strict quality control measures.

Indian Jute Industries Research Association (IJIRA) and Survecllance General superintendent (SGS) India Ltd. under MOU with IHIRA are jointly engaged in food grade jute product certification. Certified Indian food grade jute products carry a logo as a mark of quality.

JUTE MANUFACTURE

Long before jute came to occupy a prominent place amongst the textile fibres of Europe, it formed The lower qualities are, naturally, divided into fewer varieties. Each baler has his own marks, the fibres of which are guaranteed not follow that a large crop of jute will result in low prices, for the year1906-1907was not only a record one for crops, but also *olitorius.* the raw material of a large and important industry throughout the regions of Eastern Bengal. The Hindu population made the material up into cordage, paper and cloth, the chief use of the latter being in the manufacture of gunny bags. Indeed, up to1830-1840there was little or no competition with hand labour for this class of material. The process of weaving gunnies for bags and other coarse articles by these hand-loom weavers has been described as follows: "Seven sticks or chattee weaving-posts, called *land para* or warp, are fixed upon the ground, occupying the length equal to the measure of the piece to be woven, and a sufficient number of twine or thread is wound on them as warp called *land.*

The warp is taken up and removed to the weaving machine. Two pieces of wood are placed at two ends, which are tied to the *ohari* and *okher* or roller; they are made fast to the *khoti.* The *belut* or treadle is put into the warp; next to that is the *sarsul;* a thin piece of wood is laid upon the warp, called *chupari* or regulator.

There is no sley used in this, nor is a shuttle necessary; in the room of the latter a stick covered with thread called *singa* is thrown into the warp as woof, which is beaten in by a piece of plank called *beyno,* and as the cloth is woven it is wound up to the roller. Next to this is a piece of wood called *khetone,* which is used for smoothing and regulating the woof; a stick is fastened to the warp to keep the woof straight." Gunny cloth is woven of numerous qualities, according to the purpose to which it is devoted.

Some kinds are made close and dense in texture, for carrying such seed as poppy or rape and sugar; others less close are used for rice, pulses, and seeds of like size, and coarser and opener kinds again are woven for the outer cover of packages and for the sails of country boats. There is a thin close-woven cloth made and used as garments among the females of the aboriginal tribes near the foot of the Himalayas, and in various localities a cloth of pure jute or of jute mixed with cotton is used as a sheet to sleep on, as well as for wearing purposes. To indicate the variety of uses to which jute is applied, the following quotation may be cited from the official report of Hem Chunder Kerr as applying to Midnapur.

"The articles manufactured from jute are principally gunny bags, string, rope and cord, *kampa,* a net-like bag for carrying wood or hay on bullocks, *chat,* a strip of stuff for tying bales of cotton or cloth, *dola,* a swing on which infants are rocked to sleep, *shika,* a kind of hanging shelf for little earthen pots *dulina,* a floor-cloth, *beera,* a small circular stand for wooden plates used particularly in *poojahs,* painter's brush and brush for white-washing, *ghunsi,* a waist-band worn next to the skin, *gochh-dari,* a hair-band worn by women, *mukbar,* a net bag used as muzzle for cattle, *parchula,* false hair worn by players, *rakhi-bandhan,* a slender arm-band worn at the Rakhi-poornima festival; and *dhup,* small incense sticks burned at *poojahs.*"

The fibre began to receive attention in Great Britain towards the close of the 18th century, and early in the 19th century it was spun into yarn and woven into cloth in the town of Abingdon. It is claimed that this was the first British town to manufacture the material.

For years small quantities of jute were imported into Great Britain and other European countries and into America, but it was not until the year 1832 that the fibre may be said to have made any great impression in Great Britain.

The first really practical experiments with the fibre were made in this year in Chapelshade Works, Dundee, and these experiments proved to be the foundation of an enormous industry. In common with practically all new industries progress was slow for a time, but once the value of the fibre and the cloth produced from it had become known the development was more rapid. The pioneers of the work were confronted with many difficulties; most people condemned the fibre and the cloth, many warps were discarded as unfit for weaving, and any attempt to mix the fibre with flax, tow or hemp was considered a form of deception. The real cause of most of these objections was the fact that suitable machinery and methods of treatment had not been developed for preparing yarns from this useful fibre. Warden in his *Linen Trade* says: "For years after its introduction the principal spinners refused to have anything to do with jute, and cloth made of it long retained a tainted reputation. Indeed, it was not to substitute jute yarns for those made from flax in the manufacture of the coffee bagging for their East Indian possessions, that the jute trade in Dundee got a proper start. That fortunate circumstance gave an impulse to the spinning of the fibre which it never lost, and since that period its progress has been truly astonishing." The demand for this class of bagging, which is made from fine hessian yarns, is still great. These fine Rio hessian yarns form an important branch of the Dundee trade, and in some weeks during 1906 as many as 1000 bales were despatched to Brazil, besides numerous quantities to other parts of the world.

For many years Great Britain was the only European country engaged in the manufacture of jute, the great seat being Dundee. Gradually, however, the trade began to extend, and now almost every European country is partly engaged in the trade. The success of the mechanical method of spinning and weaving of jute in Dundee and district led to the introduction of textile machinery into and around Calcutta. The first mill to be run there by power was started in 1854, while by 1872 three others had been established.

In the next ten years no fewer than sixteen new mills were erected and equipped with modern machinery from Great Britain, while in 1907 there were thirty-nine mills engaged in the industry. The expansion of the Indian power trade may be gathered from the following particulars of the number of looms and spindles from 1892 to 1906. In one or two cases the number of spindles is obtained approximately by reckoning twenty spindles per loom, which is about the average for the Indian mills.

MANUFACTURE

In their general features the spinning and weaving of jute fabrics do not differ essentially as to machinery and processes from those employed in the manufacture of hemp and heavy flax goods. Owing, however, to the woody and brittle nature of the fibre, it has to undergo a preliminary treatment peculiar to itself.

The pioneers of the jute industry, who did not understand this necessity, or rather who did not know how the woody and brittle character of the fibre could be remedied, were greatly perplexed by the difficulties they had to encounter, the fibre spinning badly into a hard, rough and hairy yarn owing to the splitting and breaking of the fibre. This peculiarity of jute, coupled also with the fact that the machinery on which it was first spun, although quite suitable for the stronger and more elastic fibres for which it was designed, required certain modifications to suit it to the weaker jute, was the cause of many annoyances and failures in the early days of the trade.

The first process in the manufacture of jute is termed batching. Batch setting is the first part of this operation; it consists of selecting the different kinds or qualities of jute for any predetermined kind of yarn. The number of bales for a batch seldom exceeds twelve, indeed it is generally about six, and of these there may be three, four or even more varieties or marks. The "streaks" 1 or "heads" of jute as they come from the bale are in a hard condition in consequence of having been subjected to a high hydraulic pressure during baling; it is therefore necessary to soften them before any further process is entered.

The streaks are sometimes partly softened or crushed by means of a steam hammer during the process of opening the bale, then taken to the "strikers-up" where the different varieties are selected and hung on pins, and then taken to the jute softening machine. The more general practice, however, is to employ what is termed a "bale opener," or "jute crusher." The essential parts of one type of bale opener are three specially shaped rollers, the peri pheries of which contain a number of 1 Also in the forms "streek," "strick" or "strike," as in Chaucer, *Cant. Tales,* Prologue 676, where the Pardoner's hair is compared with a "strike of flax." The term is also used of a handful of hemp or other fibre, and is one of the many technical applications of "strike" or "streak," which etymologically are cognate words.

Thickness of material passing through the machine. The fibre is delivered by what is called the delivery cloth, and the batcher usually selects small streaks of about i 2 lb to 2 lb weight each and passes them on to the attendant or feeder of the softening machine. These small streaks are now laid as regularly as possible upon the feed-cloth of the softening machine, a general view.

The fibre passes between a series of fluted rollers, each pair of which is kept in contact by spiral springs as shown in the figure. The standard number of pairs is sixty-three, but different lengths obtain. There is also a difference in the structure. - Jute Opener. Flutes, some being straight, and others spiral, and each pair may or may not contain the same number of flutes. The springs allow the top rollers of each pair to rise as the material passes through the machine.

Advantage is taken of this slight upward and downward movement of the top rollers to automatically regulate the flow of water and oil upon the material. The apparatus for this function is placed immediately over the II th and 12th rollers of the softening machine and an idea of its construction may be gathered from. In many cases the water and oil are applied by less automatic, but equally effective, means. The main object is to see that the liquids are distributed evenly while the fibre is passing through, and to stop the supply when the machine stops or when no fibre is passing. The uniform moistening of the fibre in this machine facilitates the subsequent operations, indeed the introduction of this preliminary process (originally by hand) constituted the first important step in the practical solution of the difficulties of jute spinning. The relative quantities of oil and water depend upon the quality of the batch. Sometimes both whale and mineral oils are used, but in most cases the whale oil is omitted. About 1 to 14 gallons of oil is the usual amount given per bale of 400 lb of jute, while the quantity of water per bale varies from 3 to 7 gallons. The delivery attendants remove the streaks, give them a twist to facilitate future handling, and place Fr.3.4. - Jute Softening Machine.

Them on what are termed jute barrows. The streaks are now handed over to the cutters who cut off the roots, and finally the material is allowed to remain for twelve to twenty-four hours to allow the mixture of oil and water to thoroughly spread over the fibre.

Fig. Improved Batching Gear

When the moisture has spread sufficiently, the material is taken to the "breaker card," the first machine in the preparing department. A certain weight of jute, termed a "dollop," is laid upon the feed cloth for each revolution of the latter. The fibre, which should be arranged on the sheet as evenly as possible, is carried up by the feed cloth and passes between the feed roller and the shell on to the small knobs. Two of these rollers are supported in the same horizontal plane of the framework, while the third or top roller is kept in close contact by means of weights and springs acting on each end of the arbor. Another type of machine termed the three pair roller jute opener. The layers from the different bales are laid upon the feed cloth which carries them up to the rollers, between which the layers are crushed and partly separated.

The proximity of the weighted roller or rollers to the fixed ones depends upon the large cylinder. This cylinder, which has a high surface speed, carries part of the fibre towards the workers and strippers; the surface speed of the workers being much slower than that of the cylinder. The pins in the two rollers oppose each other, those of the workers being "back-set," and this arrangement, combined with the relative angle of the pins, and the difference in the surface speeds of the two rollers, results in part of the fibre being broken and carried round by the worker towards the stripper. Name implies, strips the fibre off the worker, and carries it round to the cylinder. The pins of the stripper and cylinder point in the same direction, but since the surface speed of the cylinder is much greater than the surface speed of the stripper, it follows that the fibre is combed between the two, and that part is carried forward by the cylinder to be reworked. The strippers and workers are in pairs, of which there may be two or more. After passing the last pair of workers and strippers the fibre is carried forward towards the doffing roller, the pins of which are back-set, and the fibre is removed from the cylinder by the doffer, from which it passes between the drawing and pressing rollers into the conductor, and finally between the delivery and pressing rollers into the sliver can. It may be mentioned that more or less breaking takes place between each pair of rollers, the pins of which are opposed, and that combing and drawing out obtains between those rollers with pins pointing in the same direction. The ratio of the surface speeds of the drawing roller and the feed roller is termed the draft: surface speed of drawing roller draft. Surface speed of feed roller In this machine the draft is usually about thirteen.

The sliver from the can of the breaker card may be wound into balls, or it may be taken direct to the finisher card. In the latter method from eight to fifteen cans are placed behind the feed rollers, and all the slivers from these cans are united before they emerge from the machine. The main difference between a breaker card and a finisher card is that the latter is fitted with finer pins, that it contains two doffing rollers, and that it usually possesses a greater number of pairs of workers and strippers - a full circular finisher card having four sets.

After the fibre has been thoroughly carded by the above machines, the cans containing the sliver from the finisher card are taken to the first drawing frame. A very common method is to let four slivers run into one sliver at the first drawing, then two slivers from the first drawing are run into one sliver at the second drawing frame. There are several types of drawing frames, *e.g.* push-bar or slide, rotary, spiral, ring, open-link or chain, the spiral being generally used for the second drawing. All, however, perform the same function, viz., combing out the fibres and thus laying them parallel, and in addition drawing out the sliver.

The designation of the machine indicates the particular method in which the gill pins are moved. These pins are much finer than those of the breaker and finisher cards, consequently the fibres are more thoroughly separated. The draft in the first drawing varies from three to five, while that in the second drawing is usually five to seven. It is easy to see that a certain amount of draft, or drawing out of the sliver, is necessary, otherwise the various doublings would cause the sliver to emerge thicker and thicker from each machine. The doublings play a very important part in the appearance of the ultimate rove and yarn, for the chief reason for doubling threads or slivers is to minimize irregularities of thickness and of colour in the material. In an ordinary case, the total doublings in jute from the breaker card to the end of the second drawing is ninety-six: 12 X 4 X 2 = 96; and if the slivers were made thinner and more of them used the ultimate result would naturally be improved.

The final preparing process is that of roving. In this operation there is no doubling of the slivers, but each sliver passes separately through the machine, from the can to the spindle, is drawn out to about eight times its length, and receives a small amount of twist to strengthen it, in order that it may be successfully wound upon the roving bobbin by the flyer. The chief piece of mechanism in the roving frame is the gearing known as the "differential motion." It works in conjunction with the disk and scroll, the cones, or the expanding pulley, to impart an intermittingly variable speed to the bobbin (each layer of the bobbin has its own particular speed which is constant for the full traverse, but each change of direction of the builder is accompanied by a quick change of speed to the bobbin).

It is essential that the bobbin should have such a motion, because the delivery of the sliver and the speed of the flyer are constant for a given size of rove, whereas the layers of rove on the bobbin increase in length as the bobbin fills. In the jute roving frame the bobbin is termed the "follower," because its revolutions per minute are fewer than those of the flyer. Each layer of rove increases the diameter of the material on the bobbin shank; hence, at the beginning of each layer, the speed of the bobbin must be increased, and kept at this increased speed for the whole traverse from top to bottom or vice versa.

Let R = the revolutions per second of the flyer; r = the revolutions per second of the bobbin; d = the diameter of bobbin shaft plus the material; L = the length of sliver delivered per second; then (R - r) d. *7r* = L.

In the above expression R, *it* and L are constant, therefore as d increases the term (R - r) must decrease; this can happen only when r is increased, that is, when the bobbin revolves quicker. It is easy to see from the above expression that if the bobbin were the "leader" its speed would have to decrease as it filled.

The builder, which receives its motion from the disk and scroll, from the cones, or from the expanding pulley, has also an intermittingly variable speed. It begins at a maximum speed when the bobbin is empty, is constant for each layer, but decreases as the bobbin fills.

The rove yarn is now ready for the spinning frame, where a further draft of about eight is given. The principles of jute spinning are similar to those of dry spinning for flax. For very heavy jute yarns the spinning frame is not used - the desired amount of twist being given at the roving frame.

The count of jute yarn is based upon the weight in pounds of 14,400 yds., such length receiving the name of "spyndle." The finest yarns weigh 22 lb to 3 lb per spyndle, but the commonest kinds are 7 Ib, 8 lb, 9 lb and io lb per spyndle.

The sizes rise in pounds up to about 20 lb, then by 2 lb up to about 50 lb per spyndle, with much larger jumps above this weight. It is not uncommon to find zoo lb to 300 lb rove yarn, while the weight occasionally reaches 450 lb per spyndle. The different sizes of yarn are extensively used in a large variety of fabrics, sometimes alone, sometimes in conjunction with other fibres, *e.g.* with worsted in the various kinds of carpets, with cotton in tapestries and household cloths, with line and tow yarns for the same fabrics and for paddings, &c., and with wool for horse clothing. The yarns are capable of being dyed brilliant colours, but, unfortunately, the colours are not very fast to light. The fibre can also be prepared to imitate human hair with remarkable closeness, and advantage of this is largely taken in making stage wigs.

MANUFACTURING OF JUTE AND RAMIE

Jute, common name for two species of woody herbs, and for the fibre of these herbs. They are cultivated in tropical countries throughout the world. They are annuals with sparsely branching stems, often growing as tall as 4 m (14 ft). The small yellow flowers, which grow singly or in pairs, give rise to capsular fruits. The fibres of the inner bark, called bast fibres or phloem fibres, are glossy and tenacious but are brittle and easily injured by moisture.Most cultivated jute is grown in alluvial soils of the river valleys of India. The herb is planted, cultivated, and within four to five months of planting, harvested by hand.

At harvest, roots and foliage are removed from the plants, and the stems, tied in bunches, are soaked in water (retted) for two to three weeks. At the end of the retting process, the fibres are easily separable from most of the remaining stem tissue. While still in water, the stems are pounded with wooden mallets. They are then rinsed, wrung thoroughly to remove any remaining nonfibrous material, and hung up to dry. The dried fibres are yellowish white, soft, and lustrous. Jute is used in the manufacture of low-grade twine, burlap, and cheap varieties of paper. Jute is a natural fibre popularly known as the golden fibre. It is one of the cheapest and the strongest of all natural fibres and considered as fibre of the future. Jute is second only to cotton in world's production of textile fibres. India, Bangladesh, China and Thailand are the leading producers of Jute. It is also produced in southwest Asia and Brazil.

The jute fibre is also known as Pat, kosta, Nalita, Bimli or Mesta (kenaf). Kenaf known as Mesta or Ambari (species Hibiscus Cannabinus) is also considered as a variety of Jute. It is cultivated in Indian subcontinent, Thailand, China and Africa. The two main types of jute, white jute (Corchorus Capsularies)and dark jute or tossa (Corchorus Olitorius) are grown in India, Bangladesh, Thailand, China, south Asian countries and Brazil. India is the largest producer of jute goods in the world, while Bangladesh is the largest cultivator of raw jute.

Fig. Jute Gill Machine

The cultivation of Jute in India is mainly confined to the eastern region states-West Bengal, Bihar, Assam, Tripura, Meghalaya, Orrissa and Uttar Pradesh. Nearly 50 percent of total raw jute production in India alone in West Bengal. In India 4000,000 families are involved in the cultivation of raw jute. There are 76 jute mills in India and nearly 1,37,679 people are employed in these mills. Several thousand other people are engaged in several jute related diversified goods. India is also self sufficient in the jute seed production. More than 90 percent of seeds are produced by the state seed corporation of Andhra Pradesh and Maharashtra.Jute, as a natural fibre, has many inherent advantages like lusture, high tensile strength, low extensibility, moderate heat and fire resistance and long staple lengths.

It is a biodegradable and eco-friendly. It has many advantage over synthetics and protect the environment and maintain the ecological balance. Jute is not only a major textile fibre but also a raw material for non traditional and value added non-textile products. Jute is used extensively in the manufacture of different types of traditional packaging fabrics, manufacturing Hessian, saking, carpet backing, mats, bags, tarpaulins, ropes and twines. Recently jute fibres are used in a wide range of diversified products: decorative fabrics, chic-saris, salwar kamizes, soft luggage's, footwear, greeting cards, molded door panels and other innumerable useful consumer products. Supported by several technological developments today jute can be used to replace expensive fibres and scare forest materials.

The production of diversified jute products, consumer products, fashion products carving out new export market. The Indian Jute Industries Research Association (IJIRA) in association with Indian Jute Industry has recently developed Hydrocarbon free jute bags - a food grade jute bags and cloths confirming to international standard specifications. These bags are used for packaging food stuffs and in great demand throughout the world. A large number of projects are being implemented by various institutions, Government of India, its special jute development programme with UN agencies as UNDP to modernise the jute industries and its all round development.

LINDEN (PLANT)

Linden (plant), also basswood, common name for a medium-size family of mostly woody flowering plants, and for its representative genus. The family contains about 48 genera and 725 species of dicot trees and shrubs, which occur mainly in tropical regions. It is important for its ornamental and timber trees and as the source of jute.

The family characteristically has alternate, simple, asymmetrical leaves. The bark is thick and fibrous. The flowers are born in complex clusters and are small, greenish, and usually scented. They commonly have five sepals (outer floral whorls), five petals (inner floral whorls), and many stamens (male floral parts) in groups of five or ten. The ovary (female floral part) is superior (borne free from other floral parts) and matures into various kinds of several-seeded fruits.

The linden genus consists of about 45 species widely distributed in north temperate areas, including valuable ornamental trees, and timber trees that in forestry are generally called basswoods. The American basswood, which grows to a height of 30 m (100 ft), yields soft, straight-grained wood used in interiors and cabinetry and as paper pulp. Fibre from its inner bark is used to make fishnets, mats, cords, and shoes. White basswood has similar characteristics, and honey from basswood flowers is particularly flavourful. European lindens, or lime trees, are widely planted as ornamentals and have been introduced into North America. Jute, used to make cordage, burlap, and sacks, is derived from the inner bark of white jute.

JUTE ENZYMES

The downgrading of Jute fibre is the result of incomplete retting of Jute plants with residual pectin-hemicellulose gum occurring on the strands of such fibre and forming barky root end. Fibres of such quality create a great hindrance in the way of producing fine yarns and fabric. Thus Poor spinning and weaving affect productivity of Jute mills in turn hamper the cost-competitiveness of the jute.So far the benefits of Enzymes in Jute are well know to most the Jute Processor but the benefits of Bacteria are yet to be thoroughly explored. During piling operation, a biochemical reaction occurs with the help of certain Inherent Bacteria and Enzymes but those are not sufficient to give the desired softening effect.

GOLDENZ-EB

Goldenz-EB is a synergistic combination of scientific blend of Enzymes and Specific Bacterias. Enzymes like Pectinase, Xylanase and hemicellulase assists by breaking the gummy substances like Lignins, Pectins, Xylans and other natural plant polymers, which forms the barky texture of the Jute.

Whereas, specific Bacteria present in Goldenz-EB utilizes the products, which are released after Enzymatic action and further continues the process of Jute Softening. This brings about ideal softening of barky tissue, in order to loosen the bondage and that helps for easy removal of cementing portion.

JUTE CULTIVATION AND CROPPING

Jute is a rainy season crop, sown from March to May according to rainfall and type of land. It is harvested from June to September depending upon whether the sowings are early or late. Attempts have been made to grow the jute plant in America, Egypt, Africa and other places, but up to the present the fibre has proved much inferior to that obtained from plants grown in India. Here the cultivation of the plant extends from the Hugli through eastern and northern Bengal. The successful cultivation of the plant demands a hot, moist climate, with a fair amount of rain. Too much rain at the beginning of the season is detrimental to the growth, while a very dry season is disastrous. The climate of eastern and northern Bengal appears to be ideal for the growth of the plant.

The quality of the fibre and the produce per acre depend in a measure on the preparation of the soil. The ground should be ploughed about four times and all weeds removed. The seed is then sown broadcast as in the case of flax. It is only within quite recent years that any attention has been paid to the selection of the seed. The following extract from *Capital* indicates the new interest taken in it.

"Jute seed experiments are being continued and the report for 1906 has been issued. The object of these experiments is, of course, to obtain a better class of jute seed by growing plants, especially for no other purpose than to obtain their seed. The agricultural department has about 300 maunds (25,000 lb) of selected seed for distribution this year. The selling price is to be Rs. io per maund.

The agricultural department of the government of Bengal are now fully alive to the importance of fostering the jute industry by showing conclusively that attention to scientific agriculture will make two maunds of jute grow where only one maund grew before. Let them go on (as they will) till all the ryots are thoroughly indoctrinated into the new system." The time of sowing extends from the middle of March to the middle of June, while the reaping, which depends upon the time of sowing and upon the weather, is performed from the end of June to the middle of October. The crop is said to be ready for gathering when the flowers appear; if gathered before, the fibre is weak, while if left until the seed is ripe, the fibre is stronger, but is coarser and lacks the characteristic lustre. The fibre is separated from the stalks by a process of retting similar to that for flax and hemp. In certain districts of Bengal it is the practice to stack the crop for a few days previous to retting in order to allow the leaves to dry and to drop off the stalks. It is stated that the colour of the fibre is darkened if the leaves are allowed to remain on during the process of retting. It is also thought that the drying of the plants before retting facilitates the separation of the fibre. Any simple operation which improves the colour of the fibre or shortens the operation of retting is worthy of consideration. The benefits to be derived from the above process, however, cannot be great, for the bundles are usually taken direct to the pools and streams.

The period necessary for the completion of the retting process varies according to the temperature and to the properties of the water, and may occupy from two days to a month. After the first few days of immersion the stalks are examined daily to test the progress of the retting. When the fibres are easily separated from the stalk, the operation is complete and the bundles should be withdrawn.

The following description of the retting of jute is taken from Royle's *Fibrous Plants of India:-* "The proper point being attained, the native operator, standing up to his middle in water, takes as many of the sticks in his hands as he can grasp, and removing a small portion of the bark from the ends next the roots, and grasping them together, he strips off the whole with a little management from end to end, without breaking either stem or fibre.

Having prepared a certain quantity into this half state, he next proceeds to wash off: this is done by taking a large handful; swinging it round his head he dashes it repeatedly against the surface of the water, drawing it through towards him, so as to wash off the impurities; then, with a dexterous throw he fans it out on the surface of the water and carefully picks off all remaining black spots. It is now wrung out so as to remove as much water as possible, and then hung up on lines prepared on the spot, to dry in the sun." The separated fibre is then made up into bundles ready for sending to one of the jute presses. The jute is carefully sorted into different qualities, and then each lot is subjected to an enormous hydraulic pressure from which it emerges in the shape of the well-known bales

Climate and Soils

Jute requires a warm and humid climate with temperature between 24° C to 37° C. Constant rain or water-logging is harmful. The new gray alluvial soil of good depth, receiving salt from annual floods, is best for jute. Flow ever jute is grown widely in sandy loams and clay loams.

Sowing of Jute

Sowing of jute in midlands and high lands starts with showers in March or April and continues till early June in the western part of the jute belt. Compost or firm yard manure, Phosphorus and Potash, Nitrogen fertilizers are used as a fertilizer. Interculturing is essential in the early stage. Pasts are also require for plant protection.

Harvesting

Jute is harvested any time between 120 days to 150 days when the flowers have been shed, early harvesting gives good healthy fibres. The plant from 8 to 12 feet high are cut with stickles at or close the ground level. In flooded land, plants are up rooted.The harvested plants are left in field for 3 days for the leaves to shed. The stems are then made up into bundles for steeping in water. Steeping is carried out immediately after harvest.

The Fibre Extraction

The jute plant's fibres lie beneath the bark and surrounded the woody central part of the stem. To extract the fibres from the stem, the process is carried out in the following stages:

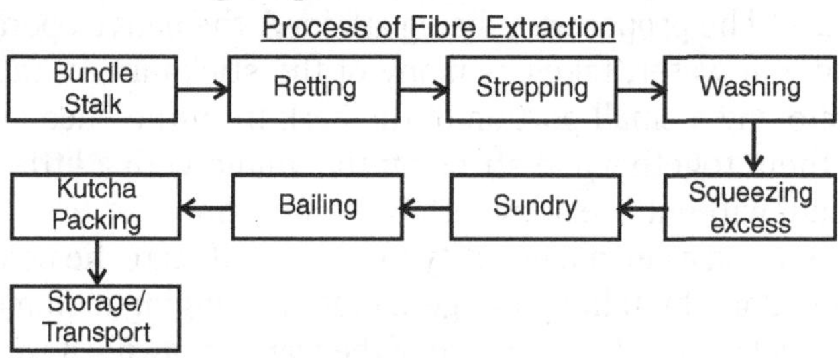

Fig. Process of Fibre Extraction

Retting

Retting is a process in which the tied bundles of jute stalks are taken to the tank by which fibres get loosened and separated from the woody stalk.

The bundles are steeped in water at least 60 cm to 100 cm depth.

The retting process is completed in 8 to 30 days, when the barks separate out easily from the stick or wood and the fibres are ready for extraction. A development in recent years is adoption of ribbon retting technology in jute growing trade of the country.

Stripping is the process of removing the fibres from the stalk after the completion of retting. Fibres are removed from the stalk by any one of the following

Methods :

- Single plants are taken and their fibres are taken off.
- Taken off a handful of stalks,breaking it in a to and fro motion in water.
- Washing the stalks first by standing in waist deep water and then stripping afterwards.

When there is a plenty of water, bundles of stalks are laid in the pond ditches or slow moving streams and left for 5-15 days under water. The bunch of stem is held in one hand and the root end tapped lightly with a mallet. After loosens the rest of fibres, fibres are extracted and washed.

WASHING AND DRYING

Extracted fibres are washed in clean water. The dark colour of fibres can be removed by dipping them in tamarind water for 15 to 20 minutes and again washed in clean water. After squeezing excess water the fibres are hang on bamboo railing for sun drying for 2-3 days.

BAILING AND PACKING

The jute fibre is graded into tops, middles, B, C and X-bottoms. Packing into Kutcha bales about 250 pounds for use in the home trade. They are transported to jute market or direct to jute mills.

JUTE TECHNOLOGY

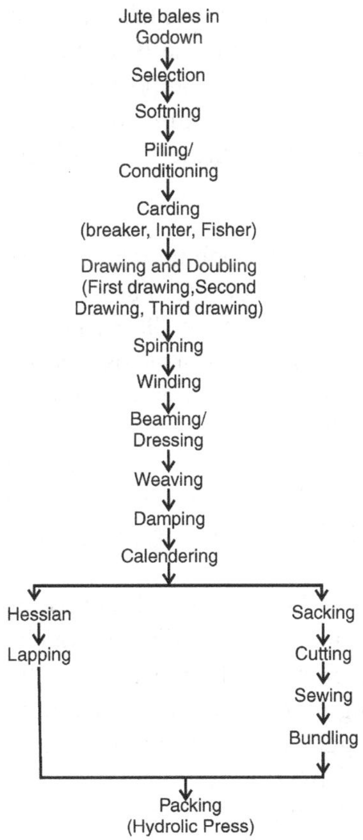

Fig. Jute Processing in Jute Mills

JUTE REINFORCED COMPOSITE TECHNOLOGY

Jute is an attractive natural fibre for use as reinforcement in composite because of its low cost, renewable nature and much lower energy requirement for processing. The scope for using jute fibres in place of the traditional glass fibres in different forms partly or fully as reinforcing agents incomposites stems from the higher specific modulus and lower specific gravity of jute (~ 40 Pa and 1.29 respectively) compared with those of glass(~ 30 GPa and 2.5 respectively).The major drawback of natural fibre reinforced composites is due to its affinity for moisture. Many experimental studies have shown that compatible coupling agents are capable of either slowing down or preventing the de- bonding process and hence moisture absorption even under severe environmental conditions, such as exposure to boiling water. Jute fibres/fabrics can be modified chemically through graft co-polymerisation and through incorporation of different resin systems by different approaches. Natural fibre composites enjoy excellent potential as wood substitutes in building industry in view of their low cost, easy availability, saving in energy and pollution free production. In order to improve upon the laboratory-industry linkages towards application development & commercialisation, the Advanced Composites Mission launched the projects on jute composites such as 'Jute-Coir Composites Boards', 'Jute-glass composite components for railway coaches', ' Thermoplastic composites based synthetic wood' and others.

Introduction

The composite technology of a polymeric matrix reinforced with man- made fibres such as glass, Kevlar, carbon etc. has come of age especially with the advances in aerospace applications since 1950s. The developments in composite material after meeting the challenges of aerospace sector have cascaded down for catering to domestic and industrial applications. Composites, the wonder material with light-weight, high strength-to-weight ratio and stiffness properties have come a long way in replacing the conventional materials like metals, woods etc. The material scientists all over the world focused their attention on natural composites reinforced with jute, sisal, coir, pineapple etc. primarily to cut down the cost of raw materials.

Eastern India has been bestowed with abundant cultivation of jute. The production of processed jute fibre in India has touched 1.44 million tonnes in 1998-99. Jute as a natural fibre has been traditionally used for making twines, ropes, cords, as packaging material in sacks & gunny bags, as carpet-backing and, more recently, as a geo-textile material. But, lately, a major share of its market has been eroded by the advent of synthetic materials, specially polypropylene. In order to save the crop from extinction and to ensure a reasonable return to the farmers, non-traditional outlets have to be explored for the fibre. One such avenue is in the area of fibre-reinforced composites.

Such composites can be used as a substitute for timber as well as in a number of less demanding applications.

Jute fibre due to its adequate tensile strength and good specific modulus enjoys the right potential for usage in composites. Jute composites can thus ensure a very effective and value-added application avenue for the natural fibre. Interest in using natural fibres as reinforcement in polymer matrices and also in certain applications as partial replacement of glass fibres has grown significantly in recent years for making low cost composite building materials. Thus, new alternative materials have emerged that could partially meet the demands of conventional materials especially wood in buildings.

NATURAL FIBRES

Jute, sisal, banana and coir, the major source of natural fibres, are grown in many parts of the world. Some of them have aspect ratios (ratio of length to diameter) > 1000 and can be woven easily. These fibres are extensively used for cordage, sacks, fishnets, matting and rope, and as filling for mattresses and cushions (e.g. rubberised coir).

Cellulosic fibres are obtained from different parts of plants, e.g. jute and ramie are obtained from the stem; sisal, banana and pineapple from the leaf; cotton from the seed; coir from the fruit, and so on.Recent reports indicate that plant-based natural fibres can very well be used as reinforcement in polymer composites, replacing to some extent more expensive and non-renewable synthetic fibres such as glass.

There are many examples of the use of cellulosic fibres in their native condition like sisal, coir jute, banana, palm, flax, cotton, and paper for reinforcement of different thermoplastic and thermosetting materials like phenol-formaldehyde, unsaturated polyester, epoxy, polyethylene, cement, natural rubber, etc.

Different geometries of these fibres, both singly and in combination with glass, have been employed for fabrication of uni-axial, bi-axial and randomly oriented composites. Amongst these various ligno-cellulosic fibres, jute contains a fairly high proportion of stiff natural cellulose. Rated fibres of jute have three principal chemical constituents, namely, a -cellulose, hemicellulose and lignin. In addition, they contain minor constituents such as fats and waxes, inorganic (mineral) matter, nitrogenous matter and traces of pigments like b - carotene and xanthophyll. As in synthetic fibre composites, the mechanical properties of the final product depend on the individual properties of the matrix, fibre and the nature of the interface between the two. Where the fibre is an agricultural one, it is possible to tailor the end properties of the composite by selection of fibres with a given chemical or morphological composition.

Several studies of fibre composition and morphology have found that cellulose content and microfibril angle tend to control the mechanical properties of cellulosic fibres. Pavithran found that higher cellulose content and lower microfibril angle resulted in higher work of fracture in impact testing.

Jute and Glass Fibres

Although the tensile strength and Young's modulus of jute are lower than those of glass fibres, the specific modulus of jute fibre is superior to that of glass and when compared on modulus per cost basis, jute is far superior. The specific strength per unit cost of jute, too, approaches that of glass. Therefore, where high strength is not a priority, jute may be used to fully or partially replace glass fibre.

The need for using jute fibres in place of the traditional glass fibre partly or fully as reinforcing agents in composites stems from its lower specific gravity (1.29) and higher specific modulus (40 Gpa) of jute compared with those of glass (2.5 & 30 Gpa respectively).

Apart from much lower cost and renewable nature of jute, much lower energy requirement for the production of jute (only 2% of that for glass) makes it attractive as a reinforcing fibre in composites. The jute composites may be used in everyday applications such as lampshades, suitcases, paperweights, helmets, shower and bath units. They are also used for covers of electrical appliances, pipes, post-boxes, roof tiles, grain storage silos, panels for partition & false ceilings, bio-gas containers, and in the construction of low cost, mobile or pre-fabricated buildings which can be used in times of natural calamities.

Effect of Moisture on Jute Fibres

There is, however, a major drawback associated with the application of jute fibres for reinforcement of resin matrices. Due to presence of hydroxy and other polar groups in various constituents of jute fibre, the moisture uptake is high (approx. 12.5% at 65% relative humidity & 20o C) by dry fibre[5].

All this leads to (i) poor wettability with resin and (ii) weak interfacial bonding between jute fibre and the relatively more hydrophobic matrices. Environmental performance of such composites is generally poor due to delimitation under humid conditions.

With increase in relative humidity up to 70%, the tenacity and Young's modulus of jute increases but beyond 70%, a decrease is observed. Thus, it is essential to pre-treat the jute fibre so that its moisture absorption is reduced and the wet ability by the resin is improved.

Modification of Jute Fibre

In order to develop composites with better mechanical properties and environmental performance, it is necessary to impart hydrophobicity To the fibres by chemical reaction with suitable coupling agents or by coating with appropriate resins. Such surface modification of jute fibre would not only decrease moisture adsorption, but would also concomitantly increase wet ability of fibres with resin and improve the interfacial bond strength, which are critical factors for obtaining better mechanical properties of composites. Modification of jute and other natural cellulose fibres can be done by following means:

Chemical means Coating with polymeric solutions and Graft copolymerisation. Jute is chemically treated with isopropyl triisostearoyl titanate (abbreviated as titanate), g - aminopropyl trimethoxy silane (abbreviated as silane), sebacoyl chloride (SC), and toluene

Diisocynate (TDI). All these reagents are expected to block the hydroxy groups of jute thus making the fibres more hydrophobic. These surface modifiers penetrate and deposit into lumens of cell wall of fibre, minimising the possible extent of moisture ingress. Polymeric coating of jute fibre with phenol-formaldehyde or resorcinol formaldehyde resins by different approaches are highly effective in enhancing the reinforcing character of jute fibre, giving as high as 20-40% improvements in flexural strength and 40-60% improvements in flexural modulus.

These modifications improve the fibre-matrix resin wettability and lead to improved bonding. Jute can be graft copolymerised with vinyl monomers such as methyl methacrylate, ethyl acrylate, styrene, vinyl acetate, acrylonitrile and acrylamide in the presence of different redox initiator systems such as vanadium - cyclohexanol, vanadium-cyclohexanone, etc. Grafting of polyacrylonitrile (10-25%) imparts 10-30% improvements in flexural strength and flexural modulus of the composites. Grafting of polymethylmethacrylate is also effective in this respect, though to a lower degree.

Jute-Polyester Composite

Polyester resin forms an intimate bond with jute fibres upto a maximum fibre: resin ratio (volume/volume) of 60:40. At this volume fraction, the Young's modulus of the composite is approximately 35 GN/m2. For higher volume fraction of fibre, the quantity of resin is insufficient to wet fibres completely. In order to overcome the poor adhesion between resin matrix and jute fibres, a multifunctional resin like polyesteramide polyol has reportedly been used as an interfacial agent.

Significant improvement in mechanical properties of jute fibre composites was observed by incorporation of polyesteramide polyol. Also, hybrid composites of glass at surface and treated jute fibre at inner core can be a good alternative. There are several types of unsaturated polyester resin - general purpose, flexible, resilient, low-shrinkage (low profile), weather resistant, chemical resistant and fire resistant varieties. T

hese polyester resins are prepared from a blend of phthalic anhydride and maleic anhydride esterified with propylene glycol to form linear polyester chains having molecular weights in the range 1000-3000.

For Curing of such unsaturated polyester resin with fibre, azo type initiators (R-N=N-R) and organic peroxides (R-O-O-R) are generally used.Fabrication of Composites Hybrid composite of glass and jute fibre can be fabricated initially by the hand lay-up technique for making the sheet-moulding compound and subsequently by using a compression-moulding machine.

10-ply hybrid laminates containing 8 inner plies of untreated/silane/titanate/ TDI treated jute fibre sandwiched between two outer plies of glass fibre (weight content of jute: 25-27%) can be made by the aforesaid process. Curing is done at 80oC under a pressure of approx. 2X105 N/m2 for a period of 90 min.Jute composites are at present being used commercially in India for applications like automobile interiors.

There are also some temporary outdoor applications like low cost housing for defence etc. However, use of jute alone as reinforcing fibre would not be suitable for high strength applications. Combination can be well suited for such applications.

Incorporation of glass with jute brings about large increase in mechanical properties of composites. Phenolic resins is one of the first synthetic resin exploited commercially for fabrication of jute-composite products mainly because of its high heat resistance, low smoke emissions excellent fire retardance properties and compatibility with jute fibres. Phenol-formaldehyde based jute composites products have been used for quite sometimes as wood & ceramic substitutes.

Fig. Jute-Polyester machine

where costs & performance have a high impact on economics, phenolic resins have been accepted in many high performance applications in composite materials. Compression moulding of composites based on jute-phenolic system has been commonly practised since few decades.

In this process, jute is impregnated with the phenolic resin by spraying process followed by drying under hot air drier. These pre-impregnated jute layers are arranged together for desired thickness and compression moulded at high pressure of 700-800 kg/m2 and at temperature of around 120-1400 C. A report from the National Institute of Research on Jute and Allied Fibre Technology (NIRJAFT), Calcutta reveals that, usually for moulded jute composites with polyester resin, the resin intake can be maximum upto 40%. Both hot press moulding and hand lay-up technique can be used for its fabrication. In the latter process, the resin take up may go upto 300-400 % on the basis of jute fibre used which is not economical. Also, it is seen that some pre-processing of jute/treatment of fibre is required so that the interface problem could be solved.

Generally, when unsaturated polyester resin is used with glass fibre, the ratio maintained is 2.5:1. Whereas, for resin with jute, the ratio maintained is 3.5-4:1. However, increase in temperature increases the productivity. Even with unsaturated polyester resin, hot condition impregnation is usually done for higher productivity.Pultrusion is a another unique process that converts primary raw materials directly into finished products, continuously and automatically, utilising most of thermoset/thermoplastic resins.

Jute, available in continuous forms such as mat, roving, tapes, yarn etc., is impregnated with resin & passed through hot die to cure the product. The speed of pultrusion ranges from 0.4 mtr. to 1 mtr. per minute depending on the complexity of the products. The loading of jute is anywhere between 50-70%. Pultruded jute composites have good electrical insulation, corrosion & high resistance properties. They find applications in roofing sheets, cable trays, doors & window frames, panelling, sections for wardrobe, partitions, etc. The characteristics of natural fibre composite boards are as follows:

Attractive natural look as it can be painted, polished or laminated at will. Water proof with minimum surface absorptio, Economical, Strong and Rigid, Environment-friendly, Can be nailed, screwed and cut sharply

INDIAN SCENARIO

Due to an occurrence of a wide variety of natural fibres in the country, Indian researchers have directed efforts for quite some time in developing innovative natural fibre composites for various applications. Development of diversified composite materials as wood substitutes is being considered an attractive solution with a view to conserve forest resources. The existing wood substitute materials such as particle/fibre board and other materials The national institutions such as National Institute of Research on Jute & Allied Fibre Technology (NIRJAFT), Indian Jute Industries Research Association (IJIRA), Central Glass & Ceramic Research Institute (CGCRI), Department of Textile Technology-IIT Delhi, Regional Research Laboratory (RRL)-Bhopal,

Institute of Jute Technology-Calcutta University, Central Building Research Institute (CBRI), Roorkee specially merit the mention for their long standing research activities. IJIRA has carried out extensive work on pre-treatment of jute fibres with acrylonitrile for improving their compatibility with thermoset plastics.

NIRJAFT has developed a whole set of novel jute & other natural fibre composite products based on hot press moulding and hand lay-up techniques.

A lot of efforts has gone into the studies on resin/fibre ratio, physical & chemical characterization of jute composites, water absorption properties etc. apart from developing products such as panels, boards, packaging material etc.The grading of raw material and its implementation for the benefit of both cultivators & industry has been one of the significant contributions of NIRJAFT. A commercially viable technology has been developed by them for the manufacture of particle boards from jute stick, which is an agro-waste.

IIT-Delhi has been quite active in developing jute-based geo-textiles for applications in prevention of soil erosion, leaching etc. CGCRI-Calcutta has worked on jute-glass hybrid components for cost reduction without sacrificing the mechanical properties.

An excellent example for commercial exploitation of jute composites has been the fabrication of automobile interiors (door panels) by Birla Jute Industries Ltd. CBRI's research activities provided new insight into the contribution of the interface to the properties of the composites.

A marked improvement in the properties of composites can occur when proper coupling agent treatment is applied to the reinforcements. The use of surface modified natural fibres in polymer matrices has been studied by several researches at CBRI.

The control of fibre-matrix adhesion is a critical factor to achieve optimal properties of composites. While the national research agencies in India have excellent scientific achievements to their credit for development of jute composites, efforts on their commercialisation have been limited so far. In order to improve upon the laboratory-industry linkages towards application development & commercialization,

The Advanced Composites Mission was conceptualised by Department of Science & Technology and Defence Research & Development Organization. The Mission mode activities are being implemented by Technology Information, Forecasting & Assessment Council (TIFAC), an autonomous organization under DST. Among a wide array of composite product development, the Mission has launched a few projects focussing on jute composites.

Characteristics

The characters by which qualities of jute are judged are colour, lustre, softness, strength, length, firmness, uniformity and absence of roots. The best qualities are of a clear whitish-yellow colour, with a fine silky lustre, soft and smooth to the touch, and fine, long and uniform in fibre. When the fibre is intended for goods in the natural colour it is essential that it should be of a light shade and uniform, but if intended for yarns which are to be dyed a dark shade, the colour is not so important.

The cultivated plant yields a fibre with a length of from 6 to so ft., but in exceptional cases it has been known to reach 14 or 15 ft. in length.

The fibre is decidedly inferior to flax and hemp in strength and tenacity; and, owing to a peculiarity in its microscopic structure, by which the walls of the separate cells composing the fibre vary much in thickness at different points, the single strands of fibre are of unequal strength. Recently prepared fibre is always stronger, more lustrous, softer and whiter than such as has been stored for some time - age and exposure rendering it brown in colour and harsh and brittle in quality.

Jute, indeed, is much more woody in texture than either flax or hemp, a circumstance which may be easily demonstrated by its behaviour under appropriate reagents; and to that fact is due the change in colour and character it undergoes on exposure to the air. The fibre bleaches with facility, up to a certain point, sufficient to enable it to take brilliant and delicate shades of dye colour, but it is with great difficulty brought to a pure white by bleaching. A very striking and remarkable fact, which has much practical interest, is its highly hygroscopic nature.

While in a dry position and atmosphere it may not possess more than 6% of moisture, under damp conditions it will absorb as much as 23 Sir G. Watt, in his *Dictionary of the Economic Products of India,* mentions the following eleven varieties of jute fibre: Serajganji, Narainganji, Desi, Deora, Uttariya, Deswal, Bakrabadi, Bhatial, Karimginji, Mirganji and Jungipuri. There are several other varieties of minor importance. The first four form the four classes into which the commercial fibre is divided, and they are commonly known as Serajgunge, Naraingunge, Daisee and Dowrah.

Serajgunge is a soft fibre, but it is superior in colour, which ranges from white to grey. Naraingunge is a strong fibre, possesses good spinning qualities, and is very suitable for good warp yarns. Its colour, which is not so high as Serajgunge, begins with a cream shade and approaches red at the roots. All the better class yarns are spun from these two kinds.

Daisee is similar to Serajgunge in softness, is of good quality and of great length; its drawback is the low colour, and hence it is not so suitable for using in natural colour. It is, however, a valuable fibre for carpet yarns, especially for dark yarns. Dowrah is a strong, harsh and low quality fibre, and is used principally for heavy wefts. Each class is subdivided according to the quality and colour of the material, and each class receives a distinctive mark called a baler's mark.

RAMIE FLOWERING

Ramie is a flowering plant which is native to Asia. It is harvested and processed to yield strong fibres, also called ramie, which are used in the production of textiles, twine, upholstery, filters, and sacking. Like flax, jute, and hemp, ramie is considered a bast fibre crop, meaning that the usable portion of the plant is found in its connective tissue structures. The plant is widely cultivated in several Asian nations, which export ramie around the world.

The scientific name for the plant is *Boehmeria nivea*, and it is also sometimes called Chinese Grass. The plant grows in the form of stalks with heart shaped leaves which sprout up from an extensive underground root system. Ramie is in the nettle family, and it has the characteristic small silvery hairs associated with nettles, although the hairs do not sting. The ramie stalk can be harvested up to six times each year in favorable cultivating conditions, although three to four crops of ramie annually are much more common.

The plants must be extensively processed to yield ramie fibre. A series of beatings, washings, and chemical treatments extracts the usable part of the plant and de-gums the fibre so that it will be usable. Once processed, ramie can be spun into thread or yarn, and it is sometimes also blended with other textile materials to make it more versatile.

Pure ramie is very strong, resistant to mold and bacteria, lustrous, and it holds its shape very well. However, ramie is also stiff, not terribly elastic, and sometimes difficult to work with because it can be very brittle. In addition, ramie does not take dye very well. All of these shortcomings make ramie more expensive than similar plant fibres, such as linen. Although the material has been used in the production of textiles for thousands of years, many producers prefer to produce it in blends rather than using it plain.

Like linen and other textiles made from woody fibre, ramie requires special care. Ideally, the fabric should be hand washed cold and not wrung or heavily pressed before being laid flat to dry. Once dried, ramie should be stored flat, and it should not be sharply creased or folded. In some cases, ramie can also be dry cleaned or machine washed on a gentle setting, and the care directions on the fabric should always be carefully followed to avoid damaging it.

CULTIVATION

Ramie is one of the oldest fibre crops, having been used for at least six thousand years, and is principally used for fabric production. It is a bast fibre, and the part used is the bark (phloem) of the vegetative stalks. Ramie is normally harvested two to three times a year but under good growing conditions can be harvested up to six times per year.[2] Unlike other bast crops, ramie requires chemical processing to de-gum the fibre.

Harvesting is done just before or soon after the beginning of flowering. It is done at this time because at this stage there is a decline in plant growth and the maximum fibre content is achieved.[2] Stems are harvested by either cutting just above the lateral roots or else bending the stem. This will enable the core to be broken and the cortex can be stripped from the plant in situ.[2]

After harvesting, stems are decorticated while the plants are fresh. If this is not done while the plants are still fresh the plants will dry out and the bark will be hard to remove. The bark ribbon is then dried as quickly as possible. This will prevent bacteria and fungi from attacking it.[2]

The dry weight of harvested stem from crops ranges from 3.4 to 4.5 t/ha/year, so a 4.5 ton crop yields 1,600 kg/ha/year of dry non-de-gummed fibre. The weight loss during de-gumming can be up to 25% giving a yield of de-gummed fibre of about 1,200 kg/ha/year. The extraction of the fibre occurs in three stages. First the cortex or bark is removed; this can be done by hand or by machine. This process is called de-cortication. Second the cortex is scraped to remove most of the outer bark, the parenchyma in the bast layer and some of the gums and pectins. Finally the residual cortex material is washed, dried, and de-gummed to extract the spinnable fibre.

HISTORY

Ramie has been around for so long that it was even used in mummy cloths in Egypt during the period 5000-3300 BC and has been grown in China for many centuries. In the study of the "Lazarus" mummy, three types of textiles were found. The outermost cloth was heavy and coarsely woven; the innermost was the lightest and most tightly woven. The outer cloth appeared to be ramie Brazil began production in the late 1930s with production peaking in 1971. Since then, production has steadily declined as a result of competition with alternative crops, such as soybeans and the important synthetic fibres. They used it in Ancient China for the farmers clothes.

PROPERTIES

Ramie is one of the strongest natural fibres. It exhibits even greater strength when wet. Ramie fibre is known especially for its ability to hold shape, reduce wrinkling, and introduce a silky lustre to the fabric appearance. It is not as durable as other fibres, and so is usually used as a blend with other fibres such as cotton or wool. It is similar to flax in absorbency, density and microscopic appearance. However it will not dye as well as cotton. Because of its high molecular crystallinity, ramie is stiff and brittle and will break if folded repeatedly in the same place; it lacks resiliency and is low in elasticity and elongation potential.[3]

USES OF RAMIE

Despite its strength, ramie has had limited acceptance for textile use. The fibre's extraction and cleaning are expensive, chiefly because of the several steps—involving scraping, pounding, heating, washing, or exposure to chemicals. Some or all are needed to separate the raw fibre from the adhesive gums or resins in which it is ensheathed. Spinning the fibre is made difficult by its brittle quality and low elasticity; and weaving is complicated by the hairy surface of the yarn, resulting from lack of cohesion between the fibres. The greater utilization of ramie depends upon the development of improved processing methods.

Ramie is used to make such products as industrial sewing thread, packing materials, fishing nets, and filter cloths. It is also made into fabrics for household furnishings (upholstery, canvas) and clothing, frequently in blends with other textile fibres (for instance when used in admixture with wool, shrinkage is reported to be greatly reduced when compared with pure wool.) Shorter fibres and waste are used in paper manufacture.

BAST (BIOLOGY)

Bast are the strong fibres in the phloem of a number of dicotyledonous plants, in particular jute, hemp, flax, ramie, kenaf, roselle hemp, etc. They support the conductive cells of the phloem and provide strength to the stem. The bast of some plants are commercially important fibre crops. Bast fibres are often called skin fibres, since the fibre is extracted from the "skin" of the plant.

USE OF BAST FIBRE

Bast fibres are processed for use in carpet yarn, rope, geotextile (netting or matting), traditional carpets, hessian or burlap, paper, sacks, etc. Bast fibres are also used in the non-woven, moulding, and composite technology industries for the manufacturing of non-woven mats and carpets, composite boards as furniture materials, automobile door panels and headliners, etc. From prehistoric times through at least the early 20th century, Bast shoes were woven from bast strips in the forest areas of Eastern Europe.

UPHOLSTERY

Upholstery refers to the materials used to stuff and cover seats, couches, chairs, and most furniture that requires such, with the exception of bed mattresses. Upholstery can include a variety of materials including leather and vinyl, foam and cotton padding.

An upholsterer will redo your upholstery when time has taken its toll on your favourite pieces of furniture. If you have a couch that has seen better days, don't throw it out just yet. You might be able to save some money by getting it recovered and restuffed. You can even change the type, pattern and colour of the upholstery when you do this, so it will seem like you got a brand new couch. Most upholsterers will bring you one or more books of sample upholstery from which you can choose. You don't have to stick with what you had before. Many times you can also pick the grade of foam to restuff it. Foam padding comes in degrees of density. Some people prefer firm padding, while others like something a little more cushioned. Once you redo the couch, your recliners might look a little tired. Not to worry! New upholstery might also do them some good! You can even get them done to match the couch. Or if you prefer, pick out chair upholstery that nicely contrasts and accents the couch.

The great advantage with redoing upholstery rather than buying new is that you can pick colours, patterns and fabrics that go perfectly with your rug, window dressings and overall colour scheme. How often have you eyed something in the showroom, only to get it home and find it clashes? With an upholstery sampler handy, you can walk around the house with the samples and place them next to the rug, drapes, and other furniture, ensuring a good match. If you already have well made furniture, replacing it with new furniture of the same quality will likely be more expensive than getting it reupholstered.

To get a quote from a local business, you will likely need to have a consultation, as the price will vary depending on the upholstery chosen, the padding and the amount of materials required. To find a local upholsterer, you can check the yellow pages. If the upholstery in your car needs replacing, you can find an automotive upholsterer to help you, though this is usually fairly pricey, perhaps because the alternative of replacing the seats is quite expensive. This might explain why seat covers are popular! Many of these businesses will also replace sagging or torn headliners. Check the automotive section of your local yellow pages to find this type of upholsterer.

Elastic

When the word elastic is used as a noun, it refers to a stretchy material or an item made with this pliable substance. Bands made of it are most frequently prepared by weaving together strands of rubber, latex, or other flexible material that can return to its original shape after being stretched or extended. Elastic is most commonly used in clothing to provide support in undergarments such as girdles or bras. Elastic is also create elastic waistbands. Strips of it are sewn into or onto a piece of fabric larger than the size that is needed. The waistband can then be stretched and the expanse of fabric will move with it, making it easy to slip such clothing on or off.

Elastic is also used to make arm and leg cuffs and to keep hoods in place rather than using ties, and can also be used as a support with strapless tops, dresses, and bras. Other uses for it include cuffs for boots, socks, or gloves in order to keep them snug. Belts too can be made from or with elastic and headbands, wristbands, and hair accessories are often made using this stretchy material. In home décor, elastic can be used for "skirts" for furniture such as vanities and it is often used to make bed skirts or bed ruffles that can be put on and taken off more easily than tradition styles. Small straps are also made from it, which reach beneath the corners of a mattress to hold sheets in place. Elastic thread, yarn, or cording can be used to make jewelry or to create embellishments for home accessories.

As an adjective, the word elastic is used in reference to people or things that are flexible. An elastic attitude or personality is one of a person who adjusts or adapts well to various circumstances. An elastic policy is one that is moderate rather than rigid and may be applied leniently. Elastic skin is supple and may return to its shape after being stretched, such as when weight is lost after a sudden gain. The terms elastic and elasticity are also used to describe some areas of economics. Simply put, elastic or flexible changes in prices can have an effect on demand for various goods or services. Even where price changes are marginal, demand may increase greatly.

Grass Ramie (Rhea)

Grass Ramie (Rhea)the product of one or more species of the genus *Boehmeria,* a member of the order Urticaceae and nearly allied to the stinging nettle genus (*Urtica*), from which, however, it differs in absence of stinging hairs. Some confusion has arisen in the use of the various terms Chinagrass, Ramie and Rhea. Two plants are concerned. One, *Boehmeria nivea,* China-grass, has been cultivated by the Chinese from very early times under the name *Tschou-ma.* The other, probably a variety of the same species, though sometimes regarded as a distinct species is the Ramie (Malay *zami*) of the Malay Islands and the Rhea of Assam.

Boehmeria nivea is a shrubby plant with the growth of the common nettle but without stinging hairs, sending up each season a number of straight shoots from a perennial underground rootstock. The long-stalked leaves recall those of the nettle in their shape and serrated margin, but their backs are 1 From the position of the words it is even not unlikely that "Pithom and Raamses" may be the addition of a redactor, and that the first author of Exod. i. It only spoke generally of store-cities. clothed with a downy substance and have a silvery appearance. The minute greenish flowers are closely arranged along a slender axis. This variety has been cultivated by the Chinese for many years, and the fibre, which is obtained from it by a tedious hand-process, has been used more or less as a substitute for silk. The variety *tenacissima* differs in its more robust habit and larger leaves, which are pale green on the face and a very much paler green on the bark. They are not downy, however, and this affords a ready means of distinction from true China-grass. *Boehmeria nivea* is sometimes found wild in India, Malaya, China and Japan, and is probably a native of further India and Malaya. China-grass and ramie are widely cultivated not only in China, Formosa, Japan, India and Malaya, but also in Queensland, Mauritius, the Cameroons, the West Indies, Brazil, Mexico and the southern states of North America, and also in south Europe. The plant, which attains a height of from 3 to 8 ft., is grown from seed, cuttings or layers, or by division of the roots. It is easy to cultivate, and thrives in almost any soil, but especially in a naturally rich, moist, light, loamy soil. For the best growth a good and equally distributed rainfall is necessary. Sudden changes of weather result in irregularities in growth, and these have a tendency to produce plants the fibres of which vary in strength. Liberal manuring is necessary, as the plant withdraws a large quantity of valuable constituents from the soil. The plants should be cut when the flower is beginning to fall and the seed to form.

It is stated that two to four crops per season may be obtained on suitable ground, each crop yielding about 4 tons of stems per acre. With only two crops per year, and a 4% yield of fibre, the resulting product would nearly reach one-third of a ton per acre. When proper attention is given to the choice of ground, and to planting, there is not much difficulty in the way of raising a good crop; the trouble arises in the extraction of the fibre. The stems when ripe are cut down, and after the leaves and small branches have been removed, the outer cover and the layers of fibre are stripped off in the form of ribbons.

These ribbons contain the bark, the fibre and a quantity of very adhesive gum. The Chinese remove this bark and as much of the gum as possible before the plant has dried. This handprocess is naturally a slow and tedious one, and many decorticators have been invented to supplant it. The action of all these decorticators is very similar. The ramie stalks are fed into the machine, and during their passage are beaten by r 2 to 20 rapidly revolving blades. These break the stalks into small pieces, and leave the bark and fibre in long ribbons. At the same time, part of the gum is squeezed out between the beaters and the anvil. Up to the present, however, these machines have not been very successful.

They usually bruise or otherwise injure the fibre, and they do not squeeze out the gum thoroughly. If the gum be allowed to dry on the ribbons it is difficult to remove it, and the chemicals employed in the degumming, if not thoroughly removed by washing, often injure the fibre to such an extent that the ultimate fabric or article is soon decomposed. If, however, the ribbons be degummed immediately, or soon after the plants are cut down, the gum will be much more easily extracted-indeed it might be possible to remove it then by boiling water or steam. The fibre cannot be expected to make much headway until the operations of decorticating and degumming are successfully carried out on or near the growing grounds; and, until a proficient decorticator is made, the fibre should be stripped by hand and the degumming operation begun immediately. By this method the least possible damage would result to the fibre, no waste material would be shipped, and a clean fibre would be placed on the market. The fibre possesses some very valuable properties; it is not only much stronger than any other known fibre, but almost equals silk in its brilliance. This latter property, however, is now challenged by mercerized cotton. It successfully resists atmospheric changes, is easily dyed and is affected but little by moisture.

On the other hand, articles manufactured from it are said to crack and break easily when sharply bent, and on account of their hairy character have not the same smart appearance as those made from flax. Although the fibre is in some cases 12 in. long, it varies considerably in length. This is one of the drawbacks in the preparing and spinning. It is impossible to make perfect yarns from fibres of various lengths; hence it is necessary either to separate the fibres into reasonable groups, or to cut them into satisfactory lengths. The latter method appears, on the whole, to be the better, and it is the method adopted by Messrs Greenwood & Batley Limited, Leeds, who make special machinery for the dressing, preparing and spinning of ramie and China-grass.

If no special machinery be employed, the length of the fibre will decide the class of machinery to be used. The fibre has been prepared and spun on flax, wool and silk-waste machinery, but it must be understood that none of these systems are really suitable for the process. A fibre with special characteristics requires special machinery for its manufacture. When so many different opinions obtain as to which existing -machinery is best adapted for the preparing and spinning of ramie, it is not surprising to find that different methods are employed in the process of manufacture. In general, however, we may say that, after decortication, the first process is that of degumming.

This is usually done by immersing the fibre in a caustic soda solution, which is then heated in a closed vessel. The fibre is laid on galvanized trays, of which as many as forty-four can be fitted in a cage, which is then placed inside the boiling keir, the lid of which is screwed down and the necessary pressure of steam admitted. After having been boiled a sufficient time to remove the gum, the material is lifted out, the alkali neutralized, and the fibre thoroughly washed to remove all traces of chemicals. The bulk of the water is removed by a hydro-extractor, and the fibre is then hung up or laid on perforated plates to dry. To facilitate the subsequent processes, the fibre is softened by passing it through a machine fitted with fluted rollers. Then follow the operations of dressing, roving, wet spinning and doubling, and finally the twisted thread is passed rapidly through a gas flame in order to remove all superfluous hairs.

In spite of the many disappointments which have been experienced in connexion with the treatment of this fibre, we are of the opinion that it will ultimately hold a good place amongst commercial fibres. It is at present spun in several European countries, but its use is still very limited. This is due, not to any imperfection of the fibre, but to its price and to the limited supply of raw material. It is at present chiefly used for gas mantles, for which it is particularly well adapted. It has also been used for paper-making, ropes, lines, nets, underwear, and for canvas and several other fabrics.

10

Textile Fibre and Power Loom

INTRODUCTION

Fibre, also spelled fibre, is a class of materials that are continuous filaments or are in discrete elongated pieces, similar to lengths of thread. They are very important in the biology of both plants and animals, for holding tissues together. Human uses for fibres are diverse. They can be spun into filaments, string or rope, used as a component of composite materials, or matted into sheets to make products such as paper or felt. Fibres are often used in the manufacture of other materials.

The strongest engineering materials are generally made as fibres, for example carbon fibre and Ultra-high-molecular-weight polyethylene. Synthetic fibres can often be produced very cheaply and in large amounts compared to natural fibres, but for clothing natural fibres can give some benefits, such as comfort, over their man-made counterparts.

TEXTILE FIBRE

A unit in which many complicated textile structures are built up is said to be textile fibre. Textile Fibre is the raw material required for the textile industry.

NATURAL FIBRES

Natural fibres include those produced by plants, animals, and geological processes. They are biodegradable over time.

They can be classified according to their origin:

- Vegetable fibres are generally based on arrangements of cellulose, often with lignin: examples include cotton, hemp, jute, flax, ramie, and sisal. Plant fibres are employed in the manufacture of paper and textile (cloth), and dietary fibre is an important component of human nutrition.
- Wood fibre, distinguished from vegetable fibre, is from tree sources. Forms include groundwood, thermo mechanical pulp (TMP) and bleached or unbleached kraft or sulfite pulps. Kraft and sulfite, also

called sulphite, refer to the type of pulping process used to remove the lignin bonding the original wood structure, thus freeing the fibres for use in paper and engineered wood products such as fibreboard.

- Animal fibres consist largely of particular proteins. Instances are spider silk, sinew, catgut, wool and hair such as cashmere, mohair and angora, fur such as sheepskin, rabbit, mink, fox, beaver, etc.
- Mineral fibres comprise asbestos. Asbestos is the only naturally occurring long mineral fibre. Short, fibre-like minerals include wollastonite, attapulgite and halloysite.

MAN-MADE FIBRES

Synthetic or man-made fibres generally come from synthetic materials such as petrochemicals. But some types of synthetic fibres are manufactured from natural cellulose, including rayon, modal, and the more recently developed Lyocell. Cellulose-based fibres are of two types, regenerated or pure cellulose such as from the cupro-ammonium process and modified cellulose such as the cellulose acetates.

Fibre classification in reinforced plastics falls into two classes:

- Short fibres, also known as discontinuous fibres, with a general aspect ratio (defined as the ratio of fibre length to diameter) between 20 to 60,
- Long fibres, also known as continuous fibres, the general aspect ratio is between 200 to 500.

CELLULOSE FIBRES

- Cellulose fibres are a subset of man-made fibres, regenerated from natural cellulose. The cellulose comes from various sources. Modal is made from beech trees, bamboo fibre is a cellulose fibre made from bamboo, seacell is made from seaweed, etc.

MINERAL FIBRES

Mineral fibres can be particular strong because they are formed with a low number of surface defects.

- Fibreglass, made from specific glass, and optical fibre, made from purified natural quartz, are also man-made fibres that come from natural raw materials, silica fibre, made from sodium silicate (water glass) and basalt fibre made from melted basalt.
- Metallic fibres can be drawn from ductile metals such as copper, gold or silver and extruded or deposited from more brittle ones, such as nickel, aluminum or iron.
- Carbon fibres are often based on oxydized and carbonised polymers, but the end product is almost pure carbon.

POLYMER FIBRES

- Polymer fibres are a subset of man-made fibres, which are based on synthetic chemicals (often from petrochemical sources) rather than arising from natural materials by a purely physical process. These fibres are made from:
  - Polyamide nylon,
  - PET or PBT polyester
  - Phenol-formaldehyde (PF)
  - Ppolyvinyl alcohol fibre (PVA)
  - Polyvinyl chloride fibre (PVC)
  - Polyolefins (PP and PE)
  - Acrylic polyesters, pure polyester PAN fibres are used to make carbon fibre by roasting them in a low oxygen environment. Traditional acrylic fibre is used more often as a synthetic replacement for wool. Carbon fibres and PF fibres are noted as two resin-based fibres that are not thermoplastic, most others can be melted.
  - Aromatic polyamids (aramids) such as Twaron, Kevlar and Nomex thermally degrade at high temperatures and do not melt. These fibres have strong bonding between polymer chains
  - Polyethylene (PE), eventually with extremely long chains/HMPE (*e.g*. Dyneema or Spectra).
  - Elastomers can even be used, *e.g*. spandex although urethane fibres are starting to replace spandex technology.
  - Polyurethane fibre
- Coextruded fibres have two distinct polymers forming the fibre, usually as a core-sheath or side-by-side. Coated fibres exist such as nickel-coated to provide static elimination, silver-coated to provide anti-bacterial properties and aluminum-coated to provide RF deflection for radar chaff. Radar chaff is actually a spool of continuous glass tow that has been aluminum coated. An aircraft-mounted high speed cutter chops it up as it spews from a moving aircraft to confuse radar signals.

MICROFIBRES

Microfibres in textiles refer to sub-denier fibre (such as polyester drawn to 0.5 dn). Denier and Detex are two measurements of fibre yield based on weight and length. If the fibre density is known you also have a fibre diameter, otherwise it is simpler to measure diameters in micrometers. Microfibres in technical fibres refer to ultra fine fibres (glass or meltblown thermoplastics) often used in filtration. Newer fibre designs include extruding fibre that splits into multiple finer fibres. Most synthetic fibres are round in cross-section, but special designs can be hollow, oval, star-shaped or trilobal. The latter design provides more optically reflective properties. Synthetic textile fibres are often crimped to provide bulk in a woven, non woven or knitted structure. Fibre surfaces can also be dull or bright. Dull surfaces reflect more light while bright tends to transmit light and make the fibre more transparent.

Very short and/or irregular fibres have been called fibrils. Natural cellulose, such as cotton or bleached kraft, show smaller fibrils jutting out and away from the main fibre structure. Once the fibre has been turned into yarn the process of making cloth is much the same for any type of fibre, be it animal or plant.

KNITTING

Hand knitting can either be done "flat" or "in the round". Flat knitting is done on a set of single point knitting needles, and the knitter goes back and forth, adding rows. In Circular knitting, or "knitting in the round", the knitter knits around a circle, creating a tube. This can be done with a set of four double pointed needles or a single circular needle. A knitted object will unravel easily if the top has not been secured. Knitted objects also stretch easily in all directions, whereas woven fabric only stretches on the bias.

CROCHETING

Crocheting differs largely from knitting in that there is only one loop, not the multitude as knitting has. Also, instead of knitting needles, a crochet hook is used.

LACE MAKING

A lace fabric is lightweight openwork fabric, patterned, with open holes in the work. The holes can be formed via removal of threads or cloth from a previously woven fabric, but more often lace is built up from a single thread and the open spaces are created as part of the lace fabric. Lace may be crocheted, or knitted.

POWER LOOM

A Northrop loom manufactured by Draper Corporation in the textile museum, Lowell, Massachusetts. The first power loom, a mechanized loom powered by a line shaft, was designed in 1784 by Edmund Cartwright and first built in 1785, it was refined over the next 47 years until a design by Kenworthy and Bullough, made the operation completely automatic.

This was known as the Lancashire Loom. By 1850 there were 250,000 in operation in England. Fifty years later came the Northrop Loom that would replenish the shuttle when it was empty and this replaced the Lancashire loom.

THE PRINCIPLES OF WEAVING

Shuttle Looms

The major components of the loom are the warp beam, heddles, harnesses, shuttle, reed and take up roll. In the loom, yarn processing includes shedding, picking, battening and taking-up operations.

- *Shedding*: Shedding is the raising of the warp yarns to form a shed through which the filling yarn, carried by the shuttle, can be inserted. The shed is the vertical space between the raised and unraised warp yarns. On the modern loom, simple and intricate shedding operations are performed automatically by the heddle or heald frame, also known as a harness. This is a rectangular frame to which a series of wires, called heddles or healds, are attached. The yarns are passed through the eye holes of the heddles, which hang vertically from the harnesses. The weave pattern determines which harness controls which warp yarns, and the number of harnesses used depends on the complexity of the weave. Two common methods of controlling the heddles are dobbies and a Jacquard Head.
- *Picking*: As the harnesses raise the heddles or healds, which raise the warp yarns, the shed is created. The filling yarn is inserted through the shed by a small carrier device called a shuttle. The shuttle is normally pointed at each end to allow passage through the shed. In a traditional shuttle loom, the filling yarn is wound onto a quill, which in turn is mounted in the shuttle. The filling yarn emerges through a hole in the shuttle as it moves across the loom. A single crossing of the shuttle from one side of the loom to the other is known as a pick. As the shuttle moves back and forth across the shed, it weaves an edge, or selvage, on each side of the fabric to prevent the fabric from raveling.
- *Battening*: As the shuttle moves across the loom laying down the fill yarn, it also passes through openings in another frame called a reed (which resembles a comb). With each picking operation, the reed presses or battens each filling yarn against the portion of the fabric that has already been formed. The point where the fabric is formed is called the fell. Conventional shuttle looms can operate at speeds of about 150 to 160 picks per minute.

With each weaving operation, the newly constructed fabric must be wound on a cloth beam. This process is called taking up. At the same time, the warp yarns must be let off or released from the warp beams. To become fully automatic, a loom needs a filling stop motion which will break the loom, if the weft thread breaks. For all this to happen, the yarn has to be prepared. The weft, or filling must be wound tightly on the correct size pirns, quills or bobbins.

Weaving happens at great speed so the yarn must be at the correct tension when it leaves the shuttle. The warp passes through the heddles which stretch it at each pick, and through the reeds which are abrasive. The warp is thus sized, that is coated with a mixture that can include china clay and flour, to give it extra strength and to act as a lubricant. It is dressed or wetted while passing through the loom. The warp, hundreds of ends of yarn rolled in parallel, comes on a wooden beam. Before weaving can commence each end must be passed through the heddles and reeds: a process known as looming.An automatic loom requires 0.125 hp to 0.5 hp to operate.

HISTORY

Rev Edmund Cartwright's invention of the power loom, and his modifications to the loom he patented in 1785 was described in his own words. It was to be forty years before his ideas were modified into a reliable automatic loom. Cartwright was not the first man to design an automatic loom, this had been done in 1678 by M. de Gennes in Paris, and again by Vaucanson in 1745, but these never developed and were forgotten. Those designs preceded John Kay's invention of the flying shuttle and they passed the shuttle through the shed using levers.

Cartwight's interest in looms was aroused by conversations in Matlock Bath on how to use the surplus yarn that was being spun on the new water frames. His first design, for a vertical loom, was made before he had observed a weaver at work. His second design adopted the handloom frame, and added mechanisms to solve known problems- instead of the warp beam he used a bobbin frame, his loom automatically *sized* or *dressed* the warp.

There was no take up reel, the clothe was delivered into a box. The *slay* or *lathe* was oscillated on primitive swords driven by a rocking rail beneath. There were shedding tappets, and cams to drive the picking. The loom only had one shaft, and the movements would have been irregular and harsh. There was a stop-motion incorporated effected by a swivel plate in the shuttle, which on dropping would engage on a hook in the shed. It was not a commercially successful machine. His ideas were licensed first by Grimshaw, of Manchester who built a small steam powered weaving factory in Manchester in 1790, the looms had to be stopped to dress the warp. The factory burnt down before anything could be learnt. A series of inventors incrementally improved all aspects of the three principle processes and the ancillary processes.

- Grimshaw 1790 Manchester- dressing the warp
- Austin 1789, 1790 -dressing the warp, 200 looms produced for Monteith of Pollockshaws 1800
- Thomas Johnson, 1803, Bredbury- dressing frame.
- Factory for 200 Steam Looms on Manchester 1806
- Two factories at Stockport 1809. One at Westhoughton, Lancashire 1809.

- William Radcliffe of Stockport 1802- improved take up mechanism
- John Todd of Burnley 1803- a heald roller and new shedding arrangements, the healds were corded to treadles actuated by cams on the second shaft.
- William Horrocks of Stockport 1803- The frame was still wooden but the lathe was pendant from the frame and operated by cams on the first shaft, the shedding was operated by cams on the second shaft, the take up motion was copied from Radcliffe.
- Peter Marsland 1806- improvements to the lathe motion to counteract poor picking
- William Cotton 1810- improvements to the letting off motion
- William Horrocks 1813 -Horrocks Loom Modifications to the lathe motion- improving on Marsland
- Peter Ewart 1813 -a use of pneumatics
- Joseph and Peter Taylor 1815 -double beat foot lathe for heavy cloths
- Paul Moody 1815- produces the first power loom in North America. Exporting a UK loom would have been illegal.
- John Capron and Sons 1820- installed the first power looms for woolens in North America at Uxbridge, Massachusetts.
- William Horrocks 1821 -a system to wet the warp and weft during use, improving the effectiveness of the sizing
- Richard Roberts 1830, Roberts Loom, These improvements were a geared take up wheel and tappets to operate multiple heddles
- Stanford, Pritchard and Wilkinson- patented a method to stop on the break of weft or warp. It was not used.
- William Dickinson of Blackburn Blackburn Loom the modern overpick loom

There now appear a series of useful improvements that are contained in patents for useless devices,

- Hornby, Kenworthy and Bullough of Blackburn 1834- the vibrating or fly reed
- John Ramsden and Richard Holt of Todmorden 1834- a new automatic weft stopping motion
- James Bullough of Blackburn 1835- improved automatic weft stopping motion and taking up and letting off arrangements
- Andrew Parkinson 1836- improved stretcher (temple).
- William Kenworthy and James Bullough 1841- trough and roller temple (became the standard), A simple stop-motion.

At this point the loom has become fully automatic. The Cartwight loom weaver could work one loom at 120-130 picks per minute- with a Kenworthy and Bullough's Lancashire Loom, a weaver can run up to six looms working at 220-260 picks per minute- thus giving 12 times more through put. The power loom is now referred to as 'a perfect machine', it produced textile of a better quality than the hand weaver for less cost. An economic success. Other improvements were the,

- James Bullough 1842 - the loose reed, which doubled the operating speed
- John Sellers 1845 - Burnley Brake, a loom brake
- Blackburn 1852- Dickinson Loom Modern overpick- or side pick using the cone and bowl that substituted the lever pick. Invented in Dickinson's mill.

LOOMS AND THE MANCHESTER CONTEXT

The development of the power loom in and around Manchester was not a coincidence. Manchester has been a centre for Fustians by 1620 and acted as a hub for other Lancashire towns, so developing a communication network with them. It was an established point of export using the meandering River Mersey, and by 1800 it had a thriving canal network, with links to the Ashton Canal, Rochdale Canal the Peak Forest Canal and Manchester Bolton and Bury Canal.

The fustian trade gave the towns a skilled workforce that was used to the complicated Dutch looms, and were perhaps accustomed to industrial discipline. While Manchester became a spinning town, the towns around were weaving towns producing cloth by the *putting out* system. The business was dominated by a few families who had the capital needed for the investment in new mills, and buy hundreds of looms.

The mills were built along the new canals so immediately had access to their markets. Spinning developed first, and until 1830 the handloom was still more important economically than the power loom when the roles reversed. Because of the economic growth of Manchester, a new industry of precision machine tool engineering was born and here were the skills needed to build the precision mechanisms of a loom.

AMERICAN CONTRIBUTIONS

Francis Cabot Lowell traveled to Great Britain from the United States in 1810 to study the textile industry there. He paid particular attention to the workings of the power loom, a device for which was not available in North America. He memorised the plans for the machine, as the British had banned the export of the new technology. On returning to Boston, he put together a group of investors, and sought the assistance of a master mechanic named Paul Moody. By 1815, Moody had built the first power loom in Waltham, Massachusetts.

In 1881 James Henry Northrop emigrated to the USA from Keighley, Yorkshire, England he worked for the Draper Corporation of Hopedale, Massachusetts. His inventions leading to an automatic "weft replenishment" loom revolutionised the weaving industry. Some 700,000 "Northrop" automatic looms were sold worldwide.

Draper's strategy was to standardise on a couple of models which it mass produced. The lighter E-model of 1909 was joined in the 1930 by the heavier X-model. Continuous fibre machines, say for rayon, which was more break-prone, needed a specialist loom. This was provided by the purchase of the Stafford Loom Co. in 1932, and using their patents a third loom the XD, was added to the range. Because of their mass production techniques they were reluctant and slow to retool for new technologies such as shuttleless looms.

DECLINE AND REINVENTION

Originally, power looms used a shuttle to throw the weft across, but in 1927 the faster and more efficient shuttleless loom came into use. Sulzer Brothers, a Swiss company had the exclusive rights to manufacture shuttleless looms in 1942, and licensed the American production to Warner and Swasey. Draper licensed the slower rapier loom. Today, advances in technology have produced a variety of looms designed to maximise production for specific types of material. The most common of these are Sulzer shuttleless weaving machines, rapier looms, air-jet looms and water-jet looms.

SOCIAL AND ECONOMIC IMPLICATIONS

The power loom reduced demand for skilled handweavers, initially causing reduced wages and unemployment. Protests followed its introduction. For example, in 1816 two thousand rioting Calton weavers tried to destroy power loom mills and stoned the workers. In the longer term, by making cloth more affordable the power loom increased demand and stimulated exports, causing a growth in industrial employment, albeit low-paid. The power loom also opened up opportunities for women mill workers. A darker side of the power loom's impact was the growth of employment of children in power loom mills.

WEAVING AND LOOM

The earliest weaving was done without a loom. In general the supporting structure of the loom is called the *frame*. It provides the means of fixing the length-wise threads, called the warp, and keeping them under tension. The warp threads are wound on a roller called the *warp beam*, and attached to the *cloth beam* which will hold the finished material. Because of the tension the warp threads are under, they need to be strong.

The thread that is woven through the warp is called the weft. The weft is threaded through the warp using a shuttle. The original *hand-loom* was limited in width by the weaver's reach, because of the need to throw the shuttle from hand to hand. The invention of the flying shuttle with its *fly cord* and *picking sticks* enabled the weaver to pass the shuttle from a *box* at either side of the loom with one hand, and across a greater width. The invention of the drop box allowed a weaver to use multiple shuttles to carry different wefts.

Alternating sets of threads are lifted by connecting them with string or wires called heddles to another bar, called the *shaft* (or *heddle bar* or *heald*). Heddles, shafts and the *couper* (lever to lift the assembly) are called the *harness* — the harness provides for mechanical operation using foot- or hand-operated *treadles*. After passing a weft thread through the warp, a reed comb is used to *beat* (compact) the woven weft.

To prepare to weave, the warp must be made. By hand this is done with the help of a warping board. The length the warp is made is about a quarter to half yard more than the amount of cloth needed. Warping boards come in a variety of shapes, from the two nearest door handles to a board with pegs on it, or a device called a warping mill that looks similar to a swift. Warping the loom, mean threading each *end* through an eye in a heddle, and then *sleying it through the reed*. The warp is *set* (verb) at X ends per inch. It then has a *sett* (noun) of X ends per inch. The weft is measured in picks per inch.

COTTAGE STAGE

Prior to the manufacturing processes being mechanized, textiles were produced in the home, and excess sold for extra money. Most cloth was made from either wool, cotton, or flax, depending on the era and location. For example, during the late medieval period, cotton became known as an imported fibre in northern Europe, without any knowledge of what it came from other than that it was a plant; noting its similarities to wool, people in the region could only imagine that cotton must be produced by plant-borne sheep.

John Mandeville, writing in 1350, stated as fact the now-preposterous belief: "There grew there [India] a wonderful tree which bore tiny lambs on the endes of its branches. These branches were so pliable that they bent down to allow the lambs to feed when they are hungry." This aspect is retained in the name for cotton in many European languages, such as German *Baumwolle*, which translates as "tree wool".

By the end of the 16th century, cotton was cultivated throughout the warmer regions in Asia and the Americas. In Roman times, wool, linen and leather clothed the European population: the cotton of India was a curiosity that only naturalists had heard of, and silk, imported along the Silk Road from China, was an extravagant luxury. The use of flax fibre in the manufacturing of cloth in Northern Europe dates back to Neolithic times.

Cloth was produced in the home, and the excess woven cloth was sold to merchants called clothiers who visited the village with their trains of pack-horses. Some of the cloth was made into clothes for people living in the same area and a large amount of cloth was exported. The process of making cloth depends slightly on the fibre being used, but there are three main steps:

preparation of fibres for spinning, spinning, and weaving or knitting. The preparation of the fibres differs the most depending on the fibre used. Flax requires retting and dressing, while wool requires carding and washing. The spinning and weaving processes are very similar between fibres though.

Spinning evolved from twisting the fibres by hand, to use of a drop spindle, to a spinning wheel. Spindles or parts of them have been found in very, very old archaeological sites; they may represent one of the earliest pieces of technology available to humankind. Was invented in India between 500 and 1000 AD it reached Europe via the Middle East in the European Middle Ages.

Weaving, done on a loom has been around for as long as spinning. There are some indications that weaving was already known in the Paleolithic. An indistinct textile impression has been found at Pavlov, Moravia. Neolithic textiles are well known from finds in pile dwellings in Switzerland. One extant fragment from the Neolithic was found in Faye at a site which dates to about 5000 BCE. There are many different types of looms, from a simple loom that dates back to the Vikings, to the standard floor loom.

HISTORY DURING THE INDUSTRIAL REVOLUTION

The key British industry at the beginning of the 18th century was the production of textiles made with wool from the large sheep-farming areas in the Midlands and across the country (created as a result of land-clearance and enclosure). Handlooms and spinning wheels were the tools of the trade of the weavers in their cottages, and this was a labour-intensive activity providing employment throughout Britain, with major centers being the West Country; Norwich and environs; and the West Riding of Yorkshire.

The export trade in woolen goods accounted for more than a quarter of British exports during most of the 18th century, doubling between 1701 and 1770. Exports of the cotton industry – centered in Lancashire – had grown tenfold during this time, but still accounted for only a tenth of the value of the woolen trade.

The textile industry grew out of the industrial revolution in the 18th Century as mass production of clothing became a mainstream industry. Starting with the flying shuttle in 1733 inventions was made to speed up the textile manufacturing process. In 1738 Lewis Paul and John Wyatt patented the Roller Spinning machine and the flyer-and-bobbin system. Lewis Paul invented a carding machine in 1748, and by 1764 the spinning jenny had also been invented.

In 1771, Richard Arkwright used waterwheels to power looms for the production of cotton cloth, his invention becoming known as the water frame. In 1784, Edmund Cartwright invented the power loom. With the spinning and weaving process now mechanized, cotton mills cropped up all over the North West of England, most notably in Manchester and its surrounding towns of Ashton-Under-Lyne, Stalybridge and Dukinfield.

Textile mills originally got their power from water wheels, and thus had to be situated along a river. With the invention of the steam engine, in the 1760s to 19th century, mills no longer needed to be along rivers.

POST INDUSTRIAL REVOLUTION

Many of the cotton mills, like the one in Lowell MA, in the US originally started with the intention of hiring local farm girls for a few years. The mill job was designed to give them a bit more money before they went back to the farm life. With the inflow of cheap labour from Ireland during the potato famine, the setup changed, as the girls became easily replaceable.

Cotton mills were full of the loud clanking of the looms, as well as lint and cotton fibre. When the mills were first built, a worker would work anywhere from one to four looms. As the design for the loom improved so that it stopped itself whenever a thread broke, and automatically refilled the shuttle, the number of machines a worker could work increased to up to 50.

Originally, power looms were shuttle-operated but in the early part of the 20th century the faster and more efficient shuttle less loom came into use. Today, advances in technology have produced a variety of looms designed to maximize production for specific types of material. The most common of these are air-jet looms and water-jet looms. Industrial looms can weave at speeds of six rows per second and faster.

By the later 20th Century, the industry in the developed world had developed a bad reputation, often involving immigrants in illegal "sweat shops" full of people working on textile manufacturing and sewing machines being paid less than minimum wages. This trend has resulted due to attempts to protect existing industries which are being challenged by developing countries in South East Asia, the Indian subcontinent and more recently, Central America.

Whilst globalization has seen the manufacturing outsourced to overseas labour markets, there has been a trend for the areas historically associated with the trade to shift focus to the more white collar associated industries of fashion design, fashion modeling and retail.

Areas historically involved heavily in the "rag trade" include London and Milan in Europe, SoHo district in New York City, the Flinders Lane and Richmond.

IN POPULAR CULTURE

- The Rag Trade, a British sitcom
- *On est au cotton*, a controversial 1970 documentary by Denys Arcand attacking management in Quebec's textile industry.
- Mentioned in the Bright Eyes (band) song, Susan Miller Rag

The industrial revolution changed the nature of work and society. Opinion varies as to the exact date when it took place but place the First Industrial

Revolution between 1750 and 1850, and the second phase or Second Industrial Revolution between 1860 and 1900. The three key drivers in these changes were textile manufacturing, iron founding and steam power. The geographical focus of Textile manufacture during the Industrial Revolution in Britain was Greater Manchester and the small towns of the Pennines and southern Lancashire. In the United States it was New England.

Prior to the 17th century, the manufacture of goods was performed on a limited scale by individual workers. This was usually on their own premises (such as weavers' cottages) – and goods were transported around the country by horse, or by river. Rivers navigations have been constructed, and some contour following canals, and, in the early 18th century, artisans were inventing ways to become more productive. Silk, Wool, Fustian, were being eclipsed by Cotton which was becoming the most important textile. This set the foundations for the changes. Historians agree that the Industrial Revolution was one of the most important events in history.

ELEMENTS OF THE INDUSTRIAL REVOLUTION

The commencement of the Industrial Revolution is closely linked to a small number of innovations, made in the second half of the 18th century:

- *Textiles*: Cotton spinning using Richard Arkwright's water frame, James Hargreaves's Spinning Jenny, and Samuel Crompton's Spinning Mule (a combination of the Spinning Jenny and the Water Frame). This was patented in 1769 and so came out of patent in 1783. The end of the patent was rapidly followed by the erection of many cotton mills. Similar technology was subsequently applied to spinning worsted yarn for various textiles and flax for linen.
- *Steam power*: The improved steam engine invented by James Watt and patented in 1775 was initially mainly used for pumping out mines, but from the 1780s was applied to power machines. This enabled rapid development of efficient semi-automated factories on a previously unimaginable scale in places where waterpower was not available.
- *Iron founding*: In the Iron industry, coke was finally applied to all stages of iron smelting, replacing charcoal. This had been achieved much earlier for lead and copper as well as for producing pig iron in a blast furnace, but the second stage in the production of bar iron depended on the use of potting and stamping (for which a patent expired in 1786) or puddling (patented by Henry Cort in 1783 and 1784).

These represent three 'leading sectors', in which there were key innovations, which allowed the economic take off by which the Industrial Revolution is usually defined. This is not to belittle many other inventions, particularly in the textile industry. Without some earlier ones, such as the spinning jenny and flying shuttle in the textile industry and the smelting of pig iron with coke, these achievements might have been impossible.

Later inventions such as the power loom and Richard Trevithick's high pressure steam engine were also important in the growing industrialisation of Britain. The application of steam engines to powering cotton mills and ironworks enabled these to be built in places that were most convenient because other resources were available, rather than where there was water to power a watermill.

In the textile sector, such mills became the model for the organisation of human labour in factories, epitomised by Cottonopolis, the name given to the vast collection of cotton mills, factories and administration offices based in Manchester. The assembly line system greatly improved efficiency, both in this and other industries. With a series of men trained to do a single task on a product, then having it moved along to the next worker, the number of finished goods also rose significantly. Also important was the 1756 rediscovery of concrete (based on hydraulic lime mortar) by the British engineer John Smeaton, which had been lost for 13 centuries.

INDUSTRY AND INVENTION

The reasons for this succession of events are interlinked. Partly through good fortune and partly through conscious effort, Britain by the early 18th century possessed the combination of skills, social needs and social resources needed for commercially successful innovation and a social system capable of sustaining the processes of rapid technological change once they had started.

Before the 1760s, textile production was a cottage industry using mainly flax and wool. In a typical house, the girls and women could make enough yarn for the man's loom. The knowledge of textile production had existed for centuries, and the manual methods had been adequate to provide enough cloth. Cotton started to be imported and the balance of demand and supply was upset.

Two systems had developed for spinning: the Simple Wheel, which used an intermittent process and the more refined Saxony wheel which drove a differential spindle and flyer with heck, in a continuous process. But neither of these wheels could produce enough thread for the looms after the invention by John Kay of the flying shuttle (which made the loom twice as productive). The first moves towards manufactories called mills were made in the spinning sector and until the 1820 cotton, wool and worsted was spun in mills, and this yarn went to outworking weavers who continues to work in their own homes.

EARLY INVENTIONS

In 1734 in Bury, Lancashire, John Kay invented the flying shuttle — one of the first of a series of inventions associated with the cotton industry. The flying shuttle increased the width of cotton cloth and speed of production of a single weaver at a loom. Resistance by workers to the perceived threat to jobs delayed the widespread introduction of this technology, even though the higher rate of production generated an increased demand for spun cotton.

START OF THE REVOLUTION

In 1761, the Duke of Bridgewater's canal connected Manchester to the coal fields of Worsley and in 1762, Matthew Boulton opened the Soho Foundry engineering works in Handsworth, Birmingham. His partnership with Scottish engineer James Watt resulted, in 1775, in the commercial production of the more efficient Watt steam engine which used a separate condensor.

In 1764, James Hargreaves is credited as inventor of the spinning jenny which multiplied the spun thread production capacity of a single worker — initially eightfold and subsequently much further. Others credit the original invention to Thomas Highs. Industrial unrest and a failure to patent the invention until 1770 forced Hargreaves from Blackburn, but his lack of protection of the idea allowed the concept to be exploited by others. As a result, there were over 20,000 Spinning Jennies in use by the time of his death. Again in 1764, Thorp Mill, the first water-powered cotton mill in the world was constructed at Royton, Lancashire, England. It was used for carding cotton.

Richard Arkwright used waterwheels to power textile machinery. His first spinning mill, Cromford Mill, Derbyshire, was built in 1771. It contained his invention the water frame. Frame is another name for the machinery for spinning or weaving. The water frame was developed from the spinning frame that Arkwright had developed with (a different) John Kay, from Warrington.

The original design was again claimed by Thomas Highs, which he claimed he had patented in 1769. Initial attempts at driving the frame had used horse power, but the innovation of using a waterwheel demanded a location with a ready supply of water, hence the mill at Cromford. This mill is preserved as part of the Derwent Valley Mills in some ways it was modelled on Matthew Boulton and John Fothergill's Soho Manufactory. Arkwright protected his investment from industrial rivals and potentially disruptive workers. He generated jobs and constructed accommodation for his workers, this led to a sizeable industrial community. Arkwright expanded his operations to other parts of the country.

Samuel Crompton of Bolton combined elements of the spinning jenny and water frame in 1779, creating the spinning mule. This mule produced a stronger thread than the water frame could. Thus in 1780, there were two viable hand operated spinning system that could be easily adapted to run by power of water. As early mules were suitable for producing yarn for use in the manufacture of muslin, and which were known as the muslin wheel or the Hall i' th' Wood (pronounced Hall-ith-wood) wheel. As with Kay and Hargreaves, Crompton was not able to exploit his invention for his own profit, and died a pauper. In 1783 a mill was built in Manchester at Shudehill, at the highest point in the city away from the river. Shudehill Mill was powered by a 30 ft diameter waterwheel. Two storage ponds were built, and the water from one passed from one to the other turning the wheel.

A steam driven pump returned the water to the higher reservoir. The steam engine was of the atmospheric type. An improvement devised by Joshua Wrigley, trialled in Chorlton-upon-Medlock used two Savery engines to supplement the river in driving on overshot waterwheel.

In 1784, Edmund Cartwright invented the power loom, and produced a prototype in the following year. His initial venture to exploit this technology failed, although his advances were recognised by others in the industry.

Others such as Robert Grimshaw (whose factory was destroyed in 1790 as part of the growing reaction against the mechanization of the industry) and Austin – developed the ideas further. In 1803, William Radcliffe invented the dressing frame which was patented under the name of Thomas Johnson which enabled power looms to operate continuously.

LATER DEVELOPMENTS

With the Cartwright Loom, the Spinning Mule and the Boulton and Watt steam engine, the pieces were in place to build a mechanised textile industry. From this point there were no new inventions, but a continuous improvement in technology as the mill-owner strove to reduce cost and improve quality. Developments in the transport infrastructure; that is the canals and after 1831 the railways facilitated the import of raw materials and export of finished cloth.

Firstly, he use of water power to drive mills was supplemented by steam driven water pumps, and then superseded completely by the steam engines. For example Samuel Greg joined his uncle's firm of textile merchants, and, on taking over the company in 1782, he sought out a site to establish a mill. Quarry was built on the River Bollin at Styal in Cheshire.

It was initially powered by a water wheel, but installed steam engines in 1810. In 1830, the average power of a mill engine was 48 hp, but Quarry Bank mill installed an new 100 hp water wheel. This was to change in 1836, when Horrocks and Nuttall, Preston took delivery of 160 hp double engine. William Fairbairn addressed the problem of line-shafting and was responsible for improving the efficiency of the mill.

In 1815 he replaced the wooden turning shafts that drove the machines at 50rpm, to wrought iron shafting working at 250 rpm, these were a third of the weight of the previous ones and absorbed less power. Secondly, in 1830, using a 1822 patent, Richard Roberts manufactured the first loom with a cast iron frame, the Roberts Loom. In 1842 James Bullough and William Kenworthy, made the Lancashire Loom. It is a semi automatic power loom. Although it is self-acting, it has to be stopped to recharge empty shuttles. It was the mainstay of the Lancashire cotton industry for a century, when the Northrop Loom invented in 1894 with an automatic weft replenishment function gained ascendancy.

Table. Number of Looms in UK

Year	1803	1820	1829	1833	1857
Looms	2400	14650	55500	100000	250000

Thirdly, also in 1830, Richard Roberts patented the first self-acting mule. Stalybridge mule spinners strike was in 1824,this stimulated research into the problem of applying power to the winding stroke of the mule. The draw while spinning had been assisted by power, but the push of the wind had been done manually by the spinner, the mule could be operated by semiskilled labour. Before 1830, the spinner would operate a partially-powered mule with a maximum of 400 spindles after, self-acting mules with upto 1300 spindles could be built.

WORKERS

Working conditions in some early British textile factories were unfavourable relative to modern standards. Children, men, and women regularly volunteered for 68-hour work weeks. Factories often were not well ventilated and became very hot in the summer. Worker health and safety regulations were non-existent. Textile factories organized workers' lives much differently from craft production. Handloom weavers worked at their own pace, with their own tools, and within their own cottages.

Factories set hours of work, and the machinery within them shaped the pace of work. Factories brought workers together within one building to work on machinery that they did not own. Factories also increased the division of labour. They narrowed the number and scope of tasks and included children and women within a common production process.

As Manchester mill owner Friedrich Engels decried, the family structure itself was "turned upside down" as women's wages undercut men's, forcing men to "sit at home" and care for children while the wife worked long hours. Factories flourished over manual craftsmanship because they had more efficient production output per worker, keeping prices down for the public, and they had much more consistent quality of product. At times, the workers rebelled against poor wages. The first major industrial action in Scotland was that of the Calton weavers in Glasgow, who went on strike for higher wages in the summer of 1787. In the ensuing disturbances, troops were called in to keep the peace and three of the weavers were killed. There was continued unrest.

In Manchester in May 1808, 15,000 protesters gathered on St George's Fields and were fired on by dragoons, with one man dying. A strike followed, but was eventually settled by a small wage increase. In the general strike of 1842, half a million workers demanded the Charter and an end to pay cuts. Again, troops were called in to keep the peace, and the strike leaders were arrested, but some of the worker demands were met.

The early textile factories employed a large share of children, but the share declined over time. In England and Scotland in 1788, two-thirds of the workers in 143 water-powered cotton mills were described as children. Sir Robert Peel, a mill owner turned reformer, promoted the 1802 Health and Morals of Apprentices Act, which was intended to prevent pauper children from working more than 12 hours a day in mills.

Children had started in the mills at around the age of four, working as scavengers under the working machinery until they were eight, they progressed to working as little piecers which they did until they were 15. During this time they worked 14 to 16 hours a day, being beaten if they fell asleep. The children were sent to the mills of Derbyshire, Yorkshire and Lancashire from the workhouses in London and other towns in the south of England.

A well documented example was that of Litton Mill. Further legislation followed. By 1835, the share of the workforce under 18 years of age in cotton mills in England and Scotland had fallen to 43%. About half of workers in Manchester and Stockport cotton factories surveyed in 1818 and 1819 had begun work at under ten years of age. Most of the adult workers in cotton factories in mid-19th century Britain were workers who had begun work as child labourers. The growth of this experienced adult factory workforce helps to account for the shift away from child labour in textile factories.

A REPRESENTATIVE EARLY SPINNING MILL 1790-1825

Cromford Mill was an early Arkwright mill and was the model for future mills The site at Cromford had year-round supply of warm water from the sough which drained water from nearby lead mines, together with another brook. It was a five storey mill. Starting in 1772,the mills ran day and night with two 12 hour shifts. It started with 200 workers, more than the locality could provide so Arkwright built housing for them nearby, one of the first manufacturers to do so. Most of the employees were women and children, the youngest being only 7 years old. Later, the minimum age was raised to 10 and the children were given 6 hours of education a week, so that they could do the record keeping their illiterate parents could not. Initially the first stage of the process was hand carding, but in 1775 he took out a second patent for a water-powered carding machine and this led to increased output. He was soon building further mills on this site and eventually employed 1,000 workers at Cromford. By the time of his death in 1792, he was the wealthiest untitled person in Britain. The gate to Cromford Mill was shut at precisely 6am and 6pm every day and any worker who failed to get through it not only lost a day's pay but was fined another day's pay. In 1779, Arkwright installed a cannon, loaded with grapeshot, just inside the factory gate, as a warning to would-be rioting textile workers, who had burned down another of his mills in Birkacre, Lancashire. The cannon was never used.

A REPRESENTATIVE MID-CENTURY SPINNING MILL 1840

Brunswick Mill, Ancoats is a cotton spinning mill in Ancoats, Manchester, Greater Manchester. It was built around 1840, part of a group of mills built along the Ashton Canal, and at that time it was one of the countries largest mill. It was built round a quadrangle, a seven storey block faced the canal. It was taken over by the Lancashire Cotton Corporation in the 1930s and passed to Courtaulds in 1964. Production finished in 1967.

The Brunswick mill was built around 1840 in one phase. The main seven storey block that faces the Ashton Canal was used for spinning. The preparation was done on the second floor and the self-acting mules with 400 spindles were arranged transversely on the floors above on the upper floor. The wings contained some spinning and ancillary processes like winding. The mill is of fireproof construction and was built by David Bellhouse, but it is suspected that William Fairbairn was involved in the design. Is was powered by a large double beam engine.

In 1850 the mill had some 276 carding machines, and 77,000 mule spindles, 20 drawing frames, fifty slubbing frames and eighty one roving frames.

A REPRESENTATIVE MID-CENTURY WEAVING SHED 1830-1860

While profiting from expertise arriving from overseas (*e.g.* Louis Paul), Britain was very protective of home-grown technology. In particular, engineers with skills in constructing the textile mills and machinery were not permitted to emigrate — particularly to the fledgeling America.

Horse power (1780–1790). The earliest cotton mills in the United States were horse powered. The first mill to use this method was the Beverly Cotton Manufactory, built in Beverly, Massachusetts. It was started August 18, 1788 by entrepreneur John Cabot and brothers. It was operated in joint by Moses Brown, Israel Thorndike, Joshua Fisher, Henry Higginson, and Deborah Higginson Cabot.

The Salem Mercury reported that in April of 1788 that the equipment for the mill was complete, consisting of a spinning jenny, a carding machine, warping machine, and other tools. That same year the mill's location was finalized and built in the rural outsets of North Beverly. The location had the presence of natural water, but it was cited the water was used for upkeep of the horses and cleaning of equipment, and not for mass-production.

Much of the internal designs of the Beverly mill were hidden due to concerns of competitors stealing designs. The beginning efforts were all researched behind closed doors, even to the point that the owners of the mill set up milling equipment on their estates to experiment with the process. There were no published articles describing exactly how their process worked in detail. Additionally, the mill's horse powered technology was quickly dwarfed by new water-powered methods.

Following the creation of the United States, an engineer who had worked as an apprentice to Arkwright's partner Jeddah Strutt evaded the ban. In 1789, Samuel Slater took his skills in designing and constructing factories to New England, and he was soon engaged in reproducing the textile mills that helped America with its own industrial revolution. Local inventions spurred this on, and in 1793 Eli Whitney invented and patented the cotton gin, which sped up the processing of raw cotton by over 350 times.

Bibliography

A.K. Dubey and Anil Kumar.: *Engineering Mechanics1*, New Age Publications, Delhi, 2008.

Alimuddin Khan.: *Basics of Classical Mechanics*, Anmol Publications, Delhi, 2008.

Anurag Saxena.: *Biomechanics in Sports*, Sports Publications, Delhi, 2011.

Archana Ruhela: *A Handbook of Livestock Management Techniques*, Oxford Book Company Publication, Delhli, 2008 .

Arvind M. Nawale: *Anita Desai's Fiction: Themes and Techniques*, B.R. Publication, Delhi, 2011.

Avtar Singh Bimbraw: *Agro-techniques for Umbelliferous Medicinal and Aromatic Plants of India*, International Book Distributing Co Publication, Delhi, 2006.

Barnes W. McCormick.: *Aerodynamics, Aeronautics, and Flight Mechanics*, Wiley India, Chennai, 2010.

D.K. Aggarwal: *Textile and Laundry in Hotel Industry*, Aman Publication, Delhi, 2006.

D.S. Malik and Pawan Kr. Bharti: *Textile Pollution*, Daya Publication, Delhi, 2010.

Eschenauer.: *Applied Structural Mechanics*, Springer Publications, Chennai, 2010.

Fomin, V.M., Kiselev, S.P., Vorozhtsov.: *Fluid Mechanics*, Jaico Publications, Delhi, 2009.

G D Arora.: *Analytical Mechanics*, Sarup Publications, Delhi, 2007.

G. S. Sawhney.: *Engineering Mechanics*, PHI Learning, Delhi, 2010.

Goutom Sharma and Julie Barooah: *Textile Tradition of Assam : Collection of Purbajyoti Sangrahalaya*, Indian Museum, 2006.

Gregory.: *Classical Mechanics*, Cambridge University Press, Delhi, 2011.

Harbhajan Singh and T.L. Singla.: *Applied Mechanics*, Abhishek Publications, Delhi, 2009.

Jatinder S. Bedi: *Economic Reforms and Textile Industry: Performance and Prospects* , Commonwealth Publication, Delhi, 2002.

Jewel Raul: *Textile Testing*, APH Publication, Delhi, 2009.

Khaja Mohtesha Muddin: *A Practical Manual of Veterinary Andrology and Reproductive Techniques*, Satish Serial Publicationg House, 2015.

M. Soundarapandian: *Textile Industry Under Globalization*, Dominant Publication, Delhi, 2004.

Rajaram Jaipuria: *Textile Legend Unravels*, Jaipuria Publication House, 2007.

Rani Kakati Hazarika and Kabita Boro: *Textile Designs of the Boros of Northeast India*, Pratibha Prakashan, Delhi, 2006.

Robin Mathew: *Textile and Apparel Industry*, Book Enclave, Delhi, 2012.

Shashi Kant Yadav.: *Textbook of Classical Mechanics*, Discovery Publications, Delhi, 2011.

Shuji Uchikawa: *Indian Textile Industry: State Policy, Liberalization and Growth*, Manohar Publication, Delhi, 2004.

Subha Gnaguly: *A Laboratory Manual on Virology and Tissue Culture Techniques*, Narendra Publication House, Delhi, 2014.

T McClurg Anderson.: *Biomechanics of Human Motion*, Sports Publications, Delhi, 2007.

T. Jeyanthi Vijayarani and Punithavathy Pandian: *Textile Exports of India : With Special Reference to Cotton Yarn*, Abhijeet Publications, Delhi, 2011.

Tanya Jain: *Textile Designing : Theory and Concept*, Yking Books, Delhi, 2012.

V.V. Vasiliev.: *Advanced Mechanics of Composite Materials*, Elsevier Publications, Delhi, 2010.

Vela Murali.: *Engineering Mechanics*, Oxford University Press, Jaipur, 2009.

Vishu Arora: *Textile Chemistry*, Abhishek Publications, Delhi, 2010.

Index